D1174916

3-26-76

# PICTORIAL ANATOMY OF THE DOGFISH

ILLUSTRATIONS & TEXT BY STEPHEN G. GILBERT

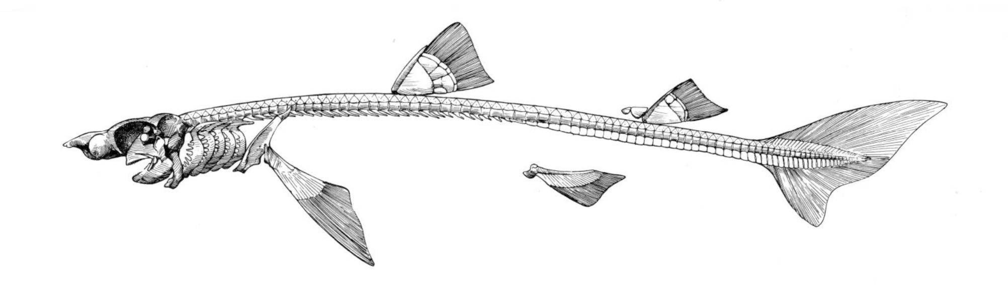

University of Washington Press Seattle and London

Second printing, 1975
Library of Congress Catalog Card Number 74-152331
ISBN 0-295-95148-6
Printed in the United States of America

# CONTENTS




2 THE SKELETON

8 THE MUSCLES

14 THE DIGESTIVE AND RESPIRATORY SYSTEMS

23 THE CIRCULATORY SYSTEM

32 THE UROGENITAL SYSTEM

38 THE NERVOUS SYSTEM

52 THE SENSE ORGANS

59 BIBLIOGRAPHY

# PICTORIAL ANATOMY OF THE DOGFISH

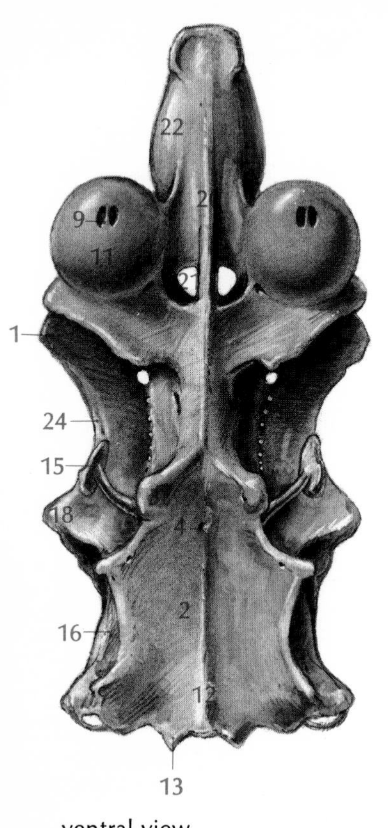


ventral view

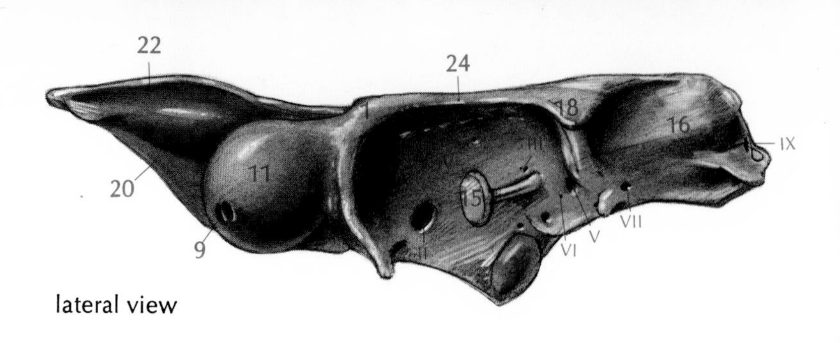


lateral view

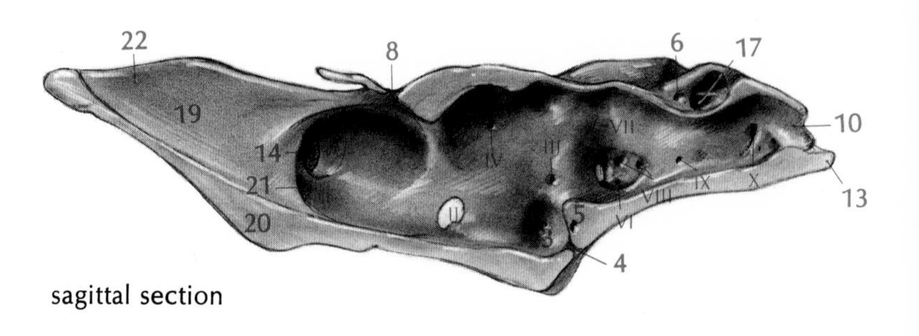


sagittal section

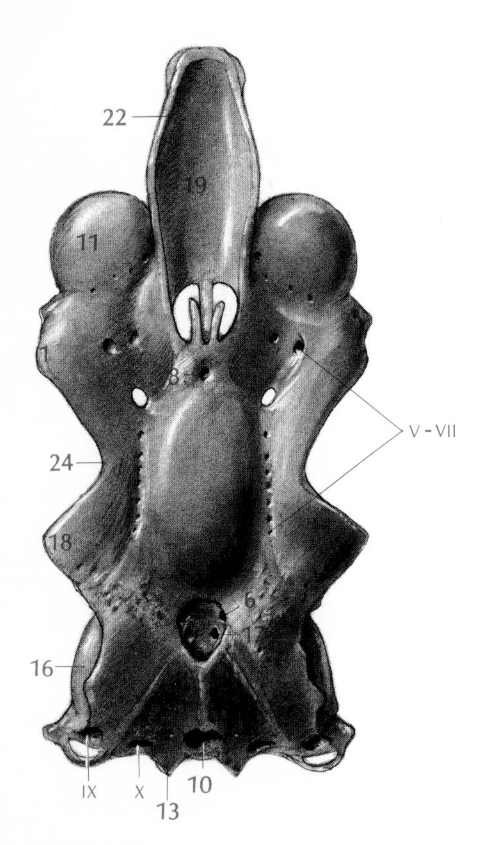


dorsal view

FIG. 1
THE CHONDROCRANIUM

1 antorbital process
2 basal plate
3 basitrabecular process
4 carotid canal
5 dorsum sellae
6 endolymphatic foramen
7 endolymphatic fossa
8 epiphysial foramen
9 external naris
10 foramen magnum
11 nasal capsule
12 notochord
13 occipital condyle
14 olfactory foramen
15 optic pedicle
16 otic capsule
17 perilymphatic foramen
18 postorbital process
19 precerebral cavity
20 rostral carina
21 rostral fenestra
22 rostrum
23 sella turcica
24 supraorbital crest

Roman numerals indicate
spinal nerve numbers.

# THE SKELETON

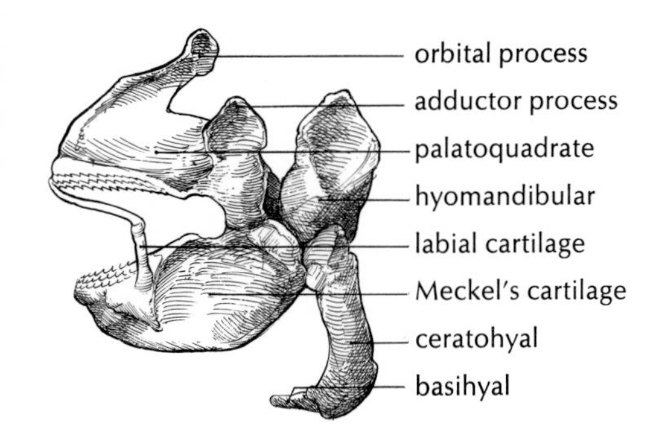


*the first and second gill arches, lateral view*

Refer to Figure 1. Examine the skull of the dogfish and note that it consists entirely of cartilage. Such a skull is termed a chondrocranium. There are no bones, and there are no divisions or sutures such as those that separate the bones of the skull in higher vertebrates.

At the anterior end of the chondrocranium is the trough-shaped rostrum. Ventral to the rostrum is a thin projection termed the rostral carina. Posterior to the rostrum on the ventral surface of the skull are the almost spherical nasal capsules. Their thin walls are easily broken and will probably not be intact in most specimens. Posteriorly the cavity of the nasal capsule communicates with the cranial cavity by way of a large opening termed the olfactory foramen; anteriorly there is a small opening termed the external naris, which is divided in two by a slender cartilage. The otic capsules, containing the inner ear, make up most of the mass of cartilage at the posterior end of the skull. Between the nasal capsule and the otic capsule is the orbit, a large depression on the lateral aspect of the chondrocranium. In life it contains the eye, which is supported by the optic pedicle, a disk supported by a slender stalk arising from the medial wall of the orbit.

Examine the dorsal surface of the chondrocranium and observe that the cavity of the rostrum (termed the precerebral cavity) communicates with the cranial cavity by way of an opening at the posterior end of the rostrum. In life this opening is closed by a membrane. Just posterior to the rostrum is the epiphysial foramen, through which the epiphysis, or pineal body, projects in life. The pineal body is the rudiment of the median third eye which was present in many primitive vertebrates. In the dorsal wall of each orbit is a series of foramina for the passage of branches of the superficial ophthalmic trunk (cranial nerves 5 and 7). In Figure 1 the foramina of the skull are identified by roman numerals indicating the number of the corresponding cranial nerves. The depression in the midline near the posterior end of the chondrocranium is the endolymphatic fossa, within which are four small openings. These openings are the paired endolymphatic foramina and the paired perilymphatic foramina, which communicate with the cavities of the inner ear in the otic capsule. The endolymphatic and perilymphatic ducts of the inner ear pass through these foramina.

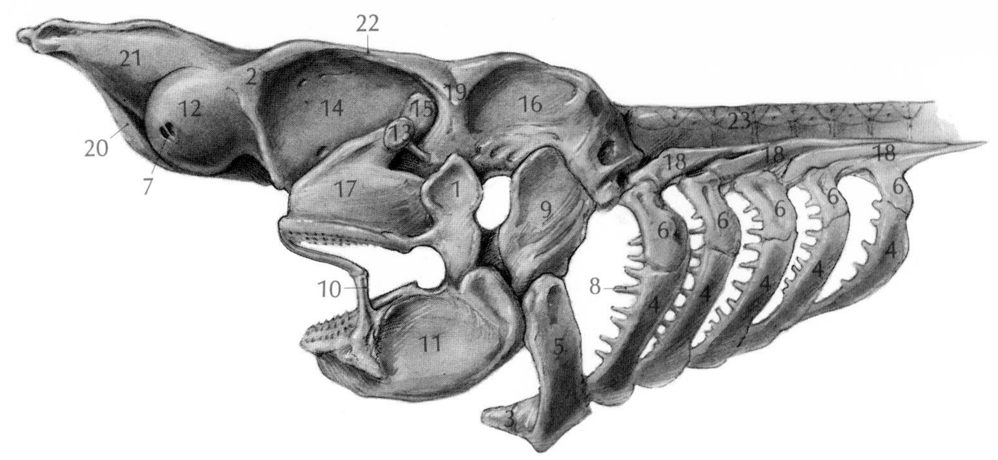


FIG. 2
THE CHONDROCRANIUM
AND GILL ARCHES

1 adductor process
2 antorbital process
3 basihyal
4 ceratobranchial
5 ceratohyal
6 epibranchial
7 external naris
8 gill raker
9 hyomandibular
10 labial cartilage
11 Meckel's cartilage
12 nasal capsule
13 optic pedicle
14 orbit
15 orbital process
16 otic capsule
17 palatoquadrate cartilage
18 pharyngobranchial
19 postorbital process
20 rostral carina
21 rostrum
22 supraorbital crest
23 vertebral column

At the posterior end of the skull observe the foramen magnum, which opens into the cranial cavity. In life the cranial cavity contains the brain, and the spinal cord passes through the foramen magnum. The paired processes ventral to the foramen magnum are the occipital condyles, which articulate with the first vertebra.

Examine the ventral surface of the skull, observing the two oval openings, termed rostral fenestrae, which communicate with the cranial cavity. The flat area between the otic capsules is the basal plate; the faint ridge in the midline of the basal plate is the notochord. At the anterior end of the notochord is the carotid canal, through which the united internal carotid arteries enter the cranial cavity. The paired lateral projections just anterior to the carotid canal are the basitrabecular processes, which aid in the suspension of the upper jaw.

Refer to Figure 2. The first gill arch forms the upper and lower jaws of the dogfish. The upper jaw consists of paired palatoquadrate cartilages; the lower jaw consists of paired Meckel's cartilages. The second gill arch, or hyoid arch, consists of two paired elements, the hyomandibular and the ceratohyal, and a single, ventrally placed, basihyal. From the palatoquadrate cartilage the orbital process extends dorsally along the medial wall of the orbit. Near its base the orbital process articulates with two facets on the lateral surface of the skull near the ventral margin of the orbit. Observe the relation of the orbital process to the basitrabecular process, which functions in the support of the jaws. The adductor process extends dorsally from the palatoquadrate near the angle of the jaws; it affords attachment for jaw muscles. The upper and lower jaws are attached to the hyoid arch by ligaments, and the hyomandibular articulates with a facet on the lateral surface of the otic capsule, to which it is attached by a ligament.

The remaining gill arches, numbered 3 through 7, typically consist of five pieces termed pharyngobranchial, epibranchial, ceratobranchial, hypobranchial, and basibranchial. In the dogfish the hypobranchials are reduced to three, and the basibranchials to two elements. The gill arches carry numerous slender lateral projections, termed gill rays, which extend into the tissue between the gill pouches, as well as short medial projections, termed gill rakers, which serve to prevent food from passing out through the gill pouches. In

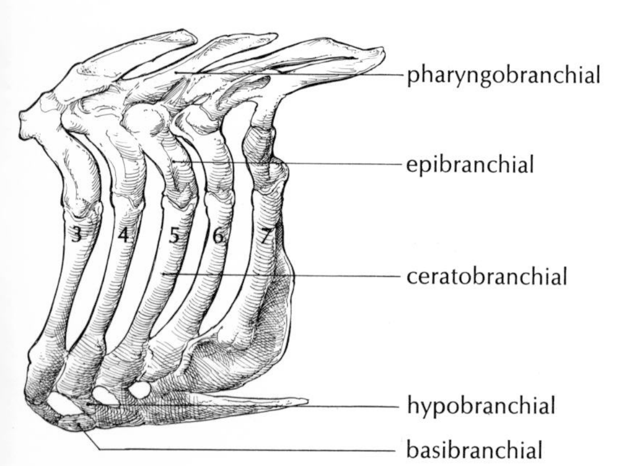


*gill arches 3-7, lateral view*

Figure 2 the gill rays are removed to expose the gill cartilages.

The vertebral column consists of a series of cylindrical vertebral bodies, or centra, dorsal to which lies a series of neural arches enclosing a neural canal within which the spinal cord lies; in the tail a similar series of haemal arches on the ventral side of the vertebral bodies encloses a haemal canal which protects the caudal artery and vein. At the tip of the neural arch is a short projection termed the neural spine; at the tip of the haemal arch is a similar projection termed the haemal spine.

In the trunk the haemal canal is absent. In its place is a series of short, laterally directed processes termed basapophyses, which are homologous with the proximal part of the haemal arches. The basapophyses give attachment to the slender ribs.

Cut through the tail just behind the posterior dorsal fin. Make a thin cross-section of the region and save it for later study. Then clean away the muscles and connective tissue from a two-inch section of the vertebral column and cut through it near the junction of two adjacent vertebrae. Observe the structures illustrated in the marginal diagram.

Now make a sagittal section of the same vertebrae and compare it with the marginal diagram. Each centrum is shaped like a spool that is hollow and concave at both ends. This type of centrum is termed amphicoelous. In sagittal section the centrum seems to consist of dorsal and ventral parts separated by diamond-shaped spaces within which the soft tissue of the notochord lies. The neural canal is seen to be composed of a series of alternating plates, the dorsal intercalary plates and the neural plates. Observe the nerve foramina in these plates. The dorsal roots of the spinal nerves exit through foramina in the dorsal intercalary plates, and the ventral roots of the spinal nerves exit through foramina in the neural plates. The haemal canal is composed of a series of elements termed haemal plates. In the lateral wall of the haemal canal are foramina through which branches of the caudal artery and vein exit.

Now remove a short section of the vertebral column from the trunk region and clean away the muscles and connective tissue. The ribs lie within a sheath of connective tissue termed the horizontal skeletogenous septum; they are slim and fragile, and the septum will have to be dissected away from them carefully. Cut through these vertebrae near the junction of two vertebral bodies and identify the structures illustrated in the marginal diagram.

Refer to Figures 3 and 4. There are two dorsal fins, termed the anterior dorsal fin and the posterior dorsal fin. Examine them on a mounted skeleton and observe that most of the flat part of the fin consists of numerous thin parallel rays. These rays are termed ceratotrichia. Near the vertebral column the ceratotrichia are supported by cartilaginous elements termed pterygiophores. There are two series of pterygiophores: the basals and the radials. The basals are larger and lie next to the vertebral column; the radials are smaller and lie between the basals and the ceratotrichia.

Examine the tail fin and observe that the ceratotrichia of the tail articulate directly with the neural and haemal arches. Also note that the vertebral column turns up into the dorsal part of the tail. This type of tail is termed heterocercal.

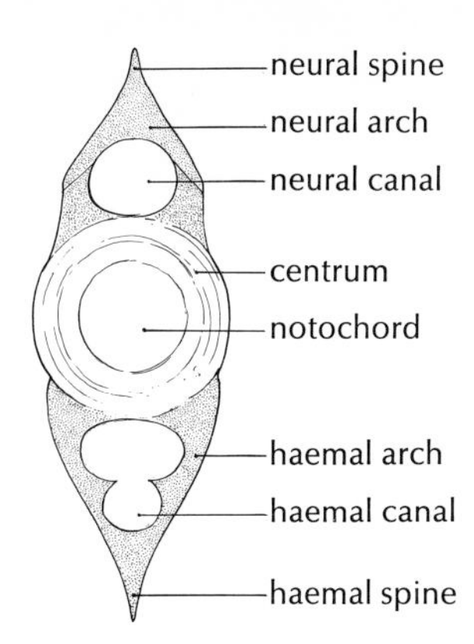

ross section of caudal vertebra

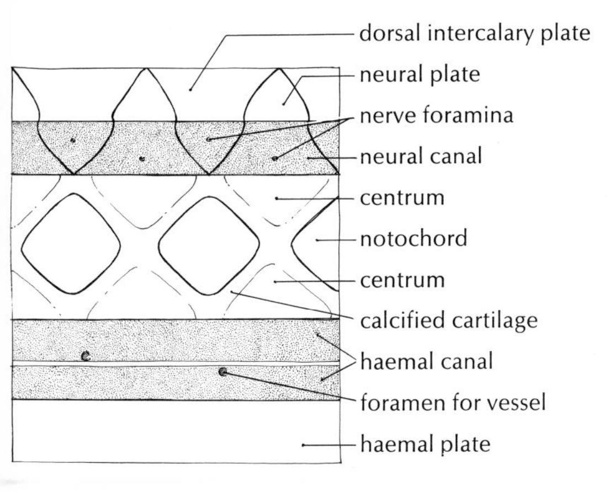

agittal section of caudal vertebra

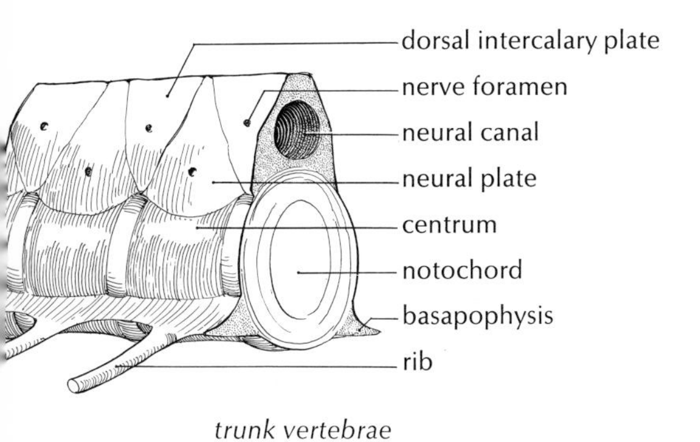

*trunk vertebrae*

ceratotrichia
spine
radials
basal

*anterior dorsal fin*

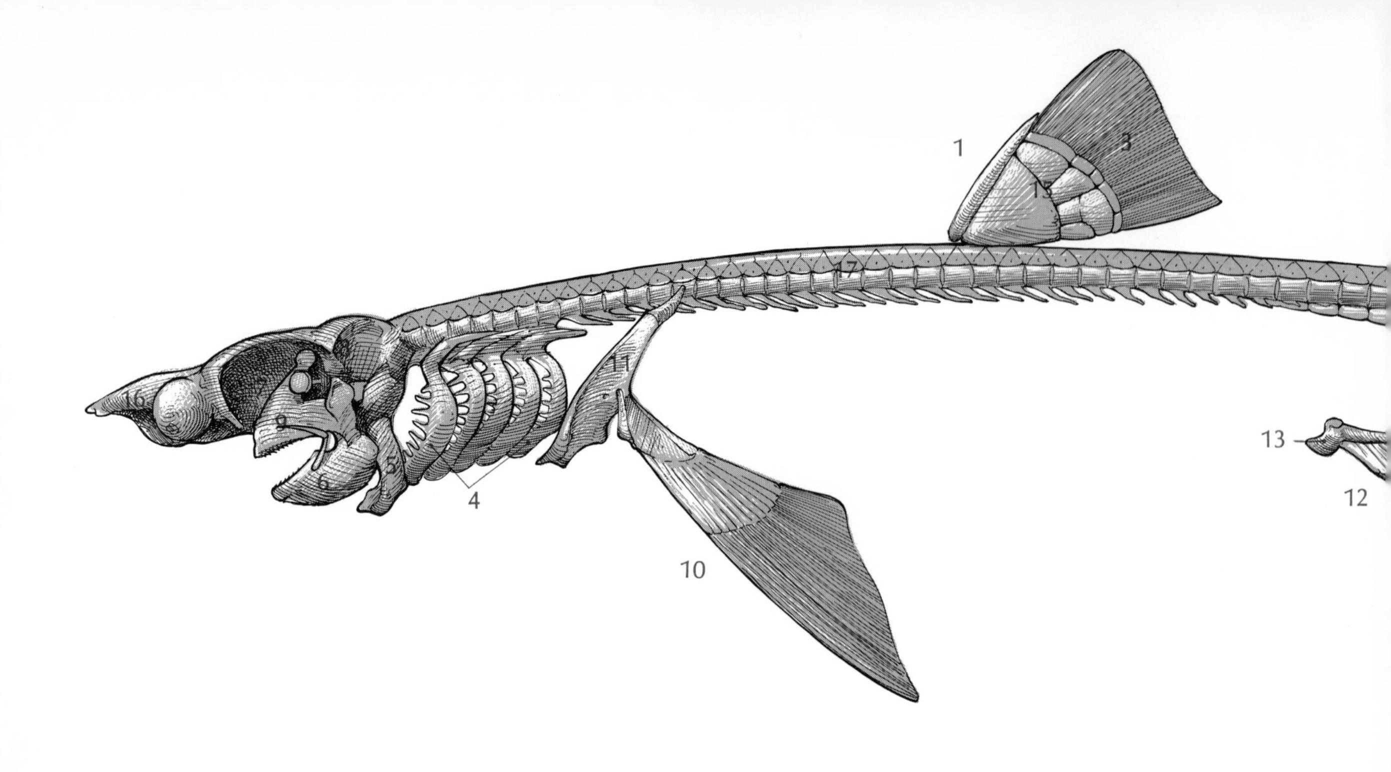


The anterior paired fins are termed the pectoral fins. The posterior paired fins are termed the pelvic fins. Basically the structure of the paired fins is similar to that of the dorsal fins. There is a series of ceratotrichia supported proximally by larger cartilages collectively termed pterygiophores, which are divided into radial and basal groups.

Examine the pectoral fin in a mounted skeleton. The radials are a series of cartilaginous rods that lie proximal to the ceratotrichia. The medial basal cartilage is termed the metapterygium; the central basal cartilage, which is the largest of the three, is termed the mesopterygium; and the lateral basal cartilage is termed the propterygium. The slender process extending dorsally above the basals is the scapular process. At the distal end of the scapular process is a separate element termed the suprascapular cartilage. Proximal to the basal cartilages is the coracoid bar, a U-shaped cartilage that encircles the ventral side of the trunk and provides attachment and support for the cartilages of the fins.

Examine the pelvic fins. In the pelvic fin there are two basals: a long metapterygium, to which most of the radials are attached, and a much smaller propterygium. Medially the basals are supported by a bar of cartilage termed the puboischiac bar. Laterally the puboischiac bar carries extensions termed iliac processes. In males a long clasper cartilage, composed of modified radials, extends posteriorly from the metapterygium and serves as a sperm-transferring organ.

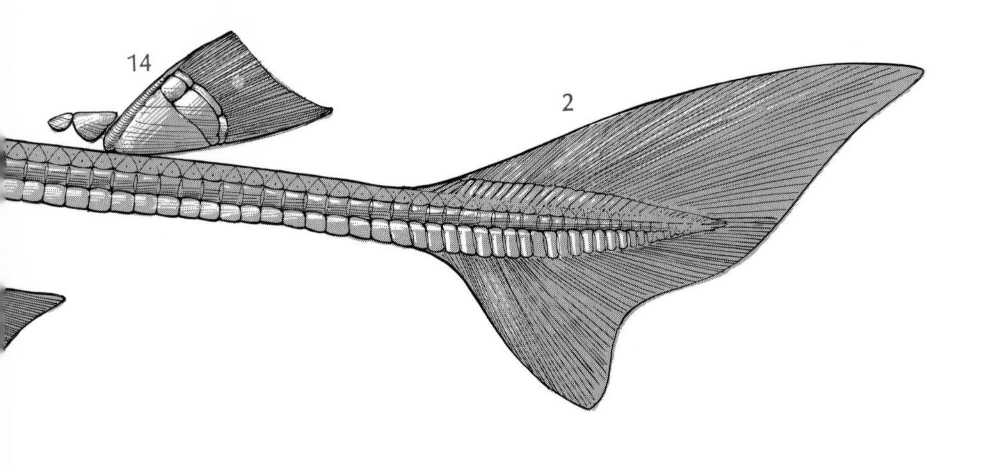


FIG. 3
THE SKELETON,
LATERAL VIEW

1 anterior dorsal fin
2 caudal fin
3 ceratotrichia (dermal fin rays)
4 gill arches 3-7
5 hyoid arch
6 Meckel's cartilage
7 orbit
8 otic capsule
9 palatoquadrate cartilage
10 pectoral fin
11 pectoral girdle
12 pelvic fin
13 pelvic girdle
14 posterior dorsal fin
15 pterygiophores (cartilaginous fin rays)
16 rostrum
17 vertebral column

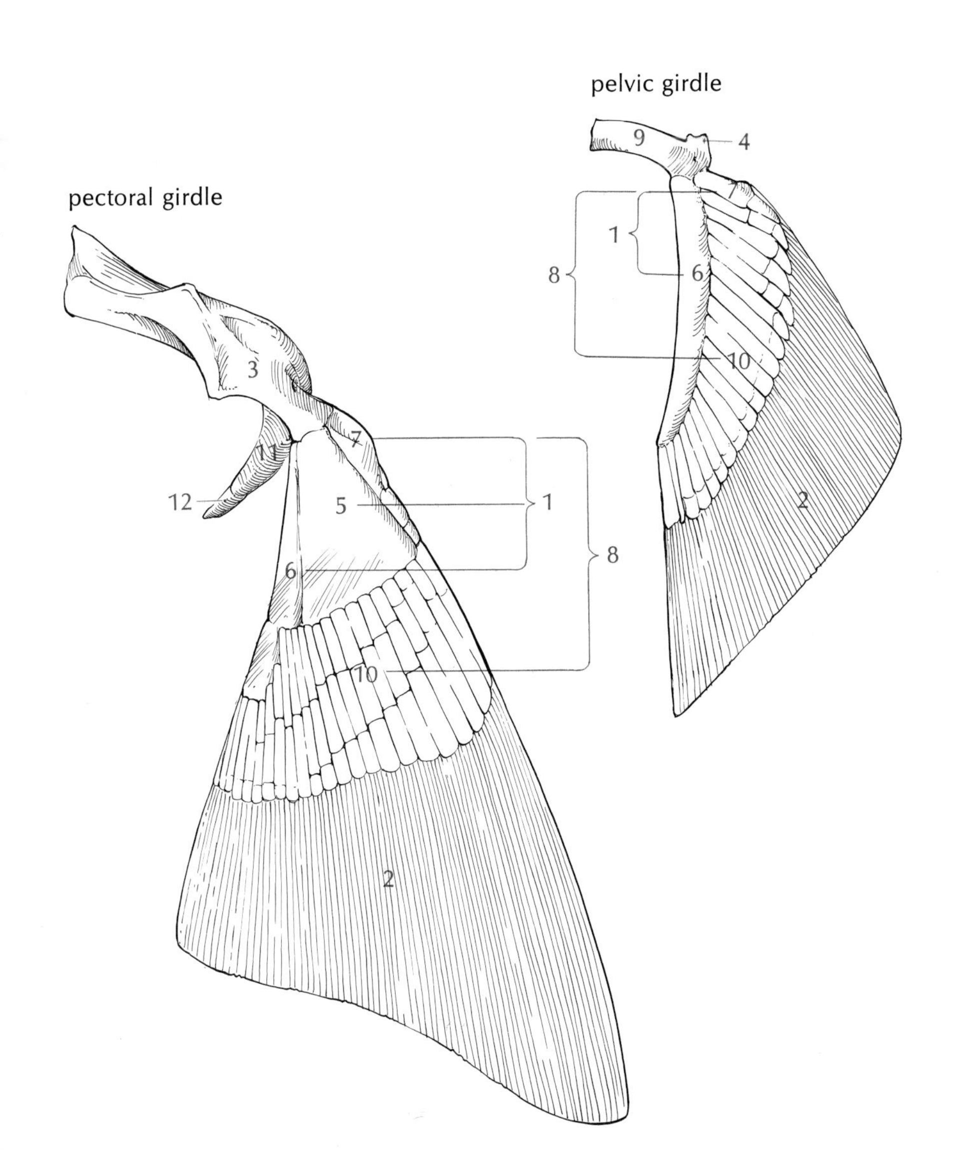


FIG. 4
THE PECTORAL AND
PELVIC FINS AND
GIRDLES, VENTRAL VIEW

1 basal cartilages
2 ceratotrichia
3 coracoid bar
4 iliac process
5 mesopterygium
6 metapterygium
7 propterygium
8 pterygiophores
9 puboischiac bar
10 radial cartilages
11 scapular process
12 suprascapular cartilage

# THE MUSCLES

dors. skeletog. septum
dors. long. bundle
lateral line
hor. skeletog. septum
lat. long. bundle
body cavity
vent. long. bundles
linea alba

*trunk muscles, cross section*

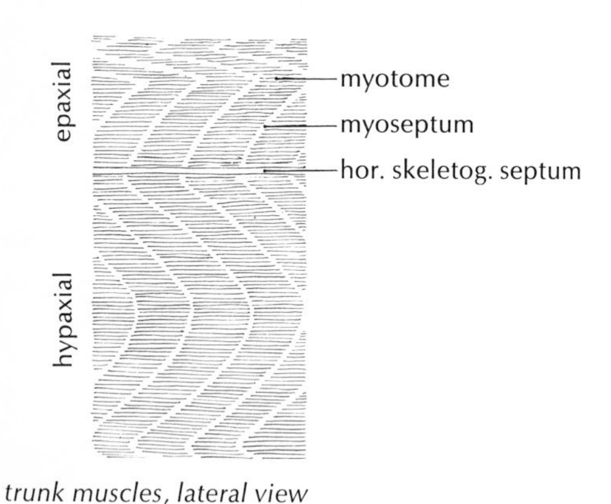


*trunk muscles, lateral view*

Starting just anterior to the posterior dorsal fin, make a superficial longitudinal cut along the dorsal side of the tail and trunk. Then make cuts along the left side of the body wall at right angles to the first cut and peel the skin away from the body by blunt dissection, keeping the underlying muscles intact as far as possible. Continue anteriorly, exposing the entire left side of the trunk.

The muscles of the tail and body wall are termed the parietal or somatic muscles. Observe that they are in the form of W-shaped segments separated by connective tissue septa. Each muscle segment is termed a myotome, and each connective tissue septum is termed a myoseptum. The muscle fibers of the myotomes run longitudinally and attach to the vertebral column or to myosepta.

The white line running the length of the body near the dorsal side is the outer edge of the horizontal skeletogenous septum, a layer of connective tissue which extends from the vertebral bodies to the skin. The parietal muscles lying above the horizontal skeletogenous septum are termed epaxial muscles; the parietal muscles lying below the horizontal skeletogenous septum are termed hypaxial muscles.

The epaxial muscles consist of two or more dorsal longitudinal bundles on each side and extend from the base of the skull to the tail. The hypaxial muscles consist of a lateral longitudinal bundle, which extends from the pectoral girdle to the tail, and two ventral longitudinal bundles, which extend from the pectoral girdle to the pelvic girdle. A septum termed the linea alba lies in the ventral midline, separating the right and left halves of the hypaxial muscles.

Skin the pelvic fins and the adjacent portions of the body and tail, and examine the muscles of the pelvic fins. Dorsal to the fin is the extensor muscle, consisting of two parts: a superficial part, which originates from the iliac process of the puboischiac bar and from the fascia of the myotomes, and a deep part, which originates from the metapterygium. Both parts insert on the radials and ceratotrichia. Cut through the superficial part of the extensor, examine its origin and insertion, and observe the deep part, which lies under it.

If your specimen is a male you will find a muscular sac, the siphon, on the ventral surface of each pelvic fin, and, medial to the pelvic fin, an oblong clasper. The clasper and the siphon are not present in

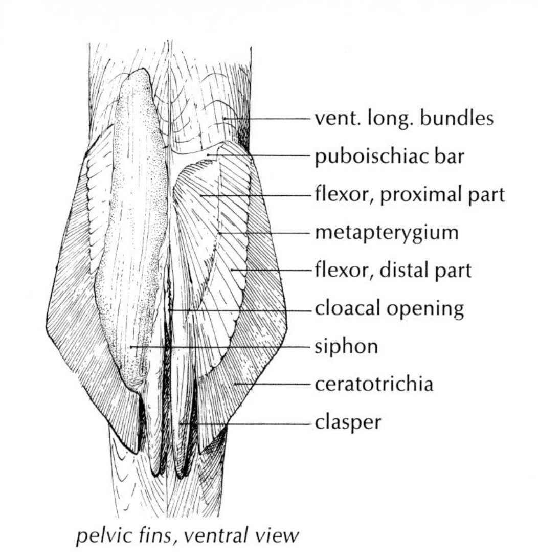


*pelvic fins, ventral view*

the female. Remove the siphon from the left fin and leave it intact on the right, as shown in the marginal illustration. The siphon and the clasper will be studied in connection with the urogenital system.

The ventral muscle of the pelvic fin is the flexor. It is divided into proximal and distal parts. The proximal part originates from the linea alba and the puboischiac bar, and inserts on the metapterygium. The distal part originates from the metapterygium and inserts on the radials and ceratotrichia. In the male the claspers are controlled by specialized portions of the pelvic fin muscles.

Refer to Figures 5, 6, and 7. Skin the head, continuing posteriorly past the pectoral girdle, and examine the muscles of the pectoral fins. Like the pelvic fins, they are controlled by dorsal extensor and ventral flexor muscles. The extensor originates from the scapular process and adjacent fascia. The flexor originates from the coracoid bar. Both the extensor and the flexor insert on the radials and ceratotrichia.

The branchial muscles, which move the gill arches and jaws, may be divided into three groups: the superficial constrictors, the levators, and the interarcuals.

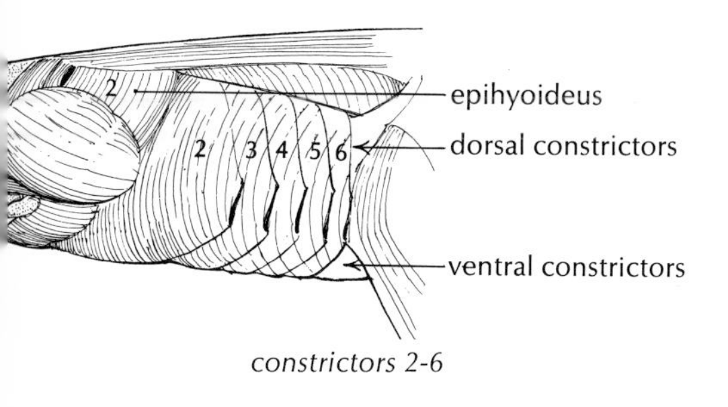


*constrictors 2-6*

The superficial constrictors cover most of the head and gill region between the eye and the sixth gill slit. They consist of dorsal constrictors, which lie dorsal to the gill slits, and ventral constrictors, which lie ventral to the gill slits. The superficial constrictors are numbered one through six; each constrictor corresponds to a gill arch of the same number. The constrictors compress the pharyngeal chamber, eject the water, and close the gill slits. The first and second dorsal constrictors elevate the upper jaw and the hyoid arch; the first and second ventral constrictors assist in opening the mouth.

Constrictors 3-6 are similar. They are attached to the connective tissue raphes which separate them and overlap each other so that each one is partially concealed by the constrictor anterior to it.

The second constrictor, corresponding to the second, or hyoid, arch, is wider than the others. It extends from the second gill slit to the angle of the jaw, and anteriorly it continues as the epihyoideus and the interhyoideus muscles. The epihyoideus lies just posterior to the spiracle. It originates from the otic capsule and surrounding fascia and inserts on the hyomandibular. The interhyoideus lies near the ventral midline just posterior to the lower jaw, concealed by the intermandibularis. It originates from the midventral raphe and inserts on the ceratohyal. Like the other constrictors, the epihyoideus and the interhyoideus compress the gill pouches. Expose the interhydoideus by cutting back the intermandibularis as illustrated in Figure 6.

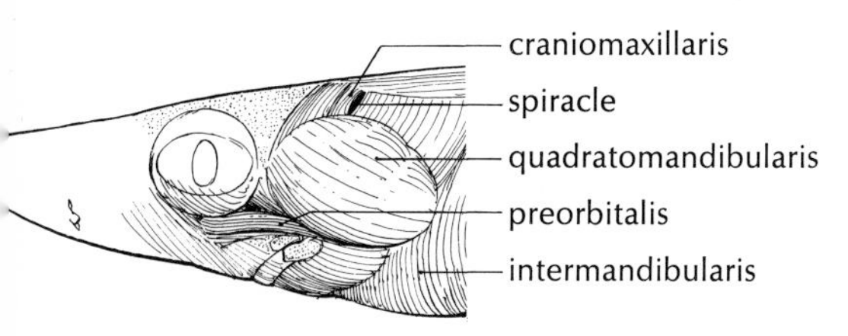


*first constrictor*

The first constrictor corresponds to the first gill arch, from which the upper and lower jaws are derived. It is represented by four separate muscles: the craniomaxillaris, preorbitalis, quadratomandibularis, and intermandibularis.

The craniomaxillaris is a small muscle which lies immediately anterior to the spiracle. Trim away as much of the chondrocranium and the wall of the spiracle as is necessary to expose the craniomaxillaris. It originates from the otic capsule and inserts on the palatoquadrate cartilage, which it elevates. Anterior to the craniomaxillaris is the levator maxillae superioris, which belongs to the levator group, described below.

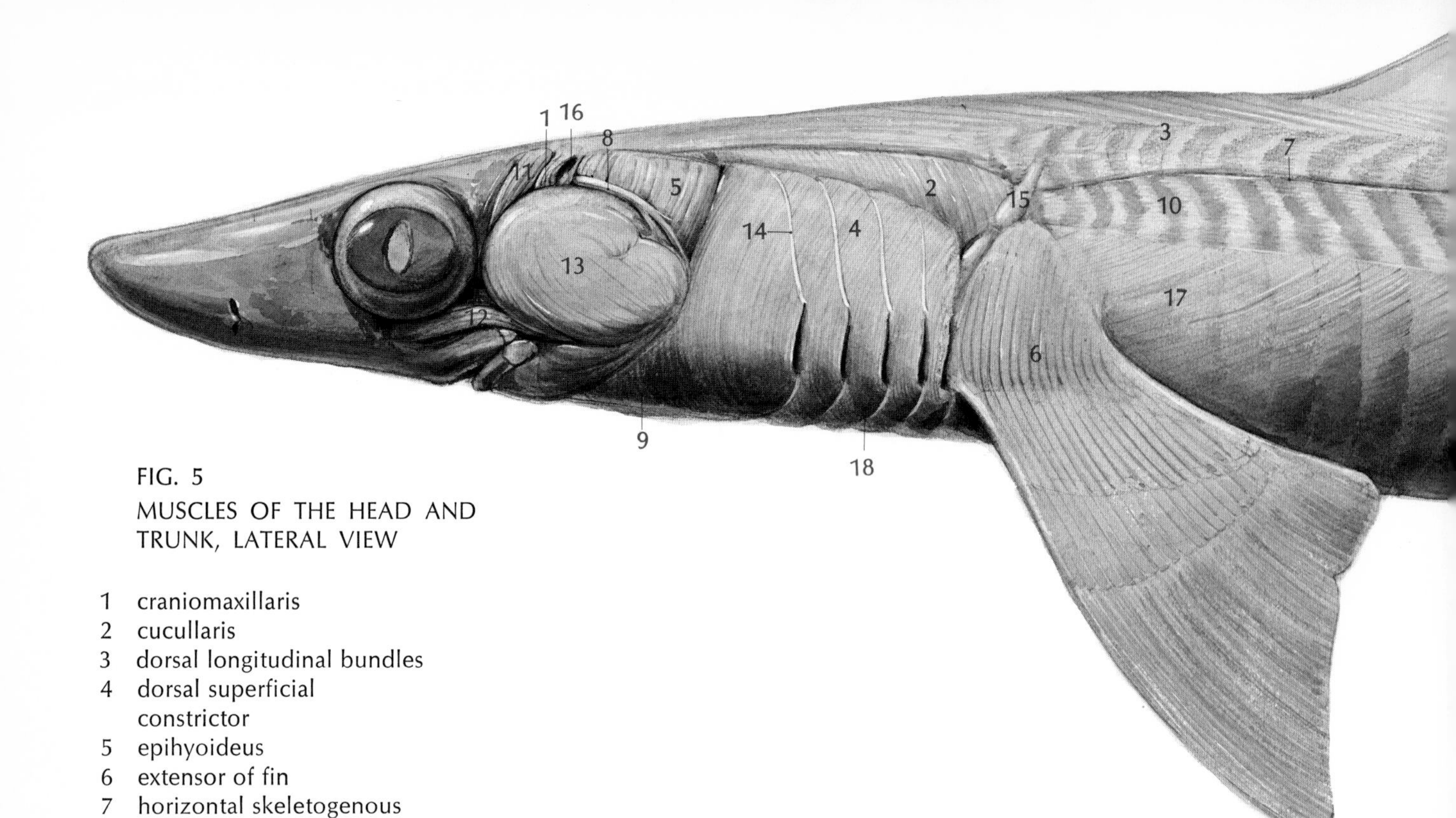


FIG. 5
MUSCLES OF THE HEAD AND TRUNK, LATERAL VIEW

1 craniomaxillaris
2 cucullaris
3 dorsal longitudinal bundles
4 dorsal superficial constrictor
5 epihyoideus
6 extensor of fin
7 horizontal skeletogenous septum
8 hyomandibular trunk of seventh cranial nerve
9 intermandibularis
10 lateral longitudinal bundle
11 levator maxillae superioris
12 preorbitalis
13 quadratomandibularis
14 raphe
15 scapular process
16 spiracle
17 ventral longitudinal bundles
18 ventral superficial constrictor

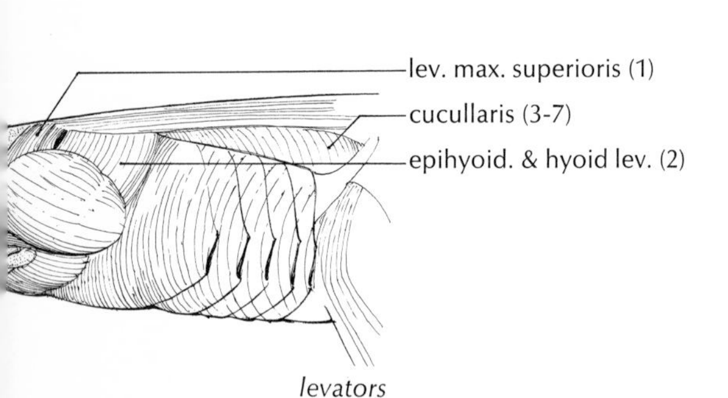


*levators*

The preorbitalis is a small muscle which lies between the upper jaw and the eye. Dissect deeply and cut away lateral portions of the upper jaw to expose it. The preorbitalis originates from the midventral surface of the chondrocranium, and its fibers insert on those of the quadratomandibularis, the large muscle at the angle of the jaw.

The quadratomandibularis originates from the posterior part of the palatoquadrate cartilage and inserts on Meckel's cartilage. Both the preorbitalis and the quadratomandibularis close the mouth.

The ventral part of the first constrictor is the intermandibularis, a broad muscle lying posterior to the mouth. It originates from the midventral raphe and inserts on Meckel's cartilage and on the fascia of the quadratomandibularis. It raises the floor of the mouth and forces water out of the gill slits.

The levator group consists of three muscles: the levator maxillae superioris, the hyoid levator, and the cucullaris.

The levator maxillae superioris, which represents the levator of the first gill arch, lies anterior to the craniomaxillaris. It originates from the otic capsule and inserts on the palatoquadrate cartilage next to the quadratomandibularis, acting to raise the upper jaw.

The second, or hyoid, levator lies deep to the epihyoideus, with which it is fused. The two muscles cannot be readily separated. The hyoid levator raises the hyoid arch (second gill arch).

The cucullaris represents the levators of gill arches 3-7. It is a long triangular muscle which lies dorsal to the constrictors and extends from the pectoral girdle anteriorly to the epihyoideus. It originates from the fascia of the dorsal longitudinal bundles and inserts on the epibranchial cartilage of the last gill arch and on the scapular process of the pectoral girdle. It elevates the gill arches and the pectoral girdle.

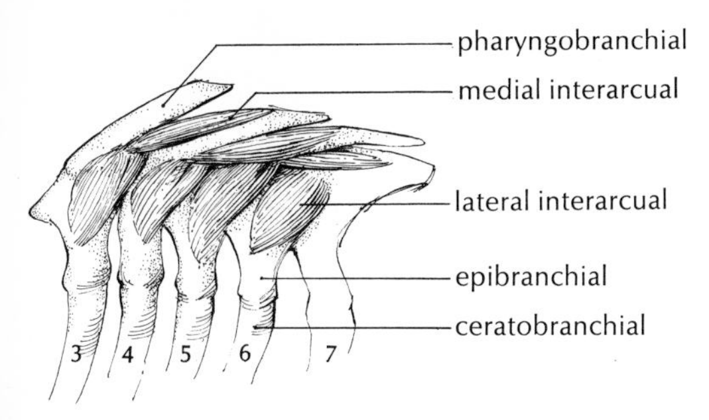


*the interarcual muscles*

The interarcuals are small muscles which act upon the gill arch cartilages to expand the pharynx. They consist of two series: the medial interarcuals, which extend between adjacent pharyngobranchial cartilages, and the lateral interarcuals, which extend from the pharyngobranchials to the corresponding epibranchials. In order to expose the interarcuals, it is necessary to destroy portions of the gills and related parts of the circulatory system; these structures should be kept intact for later study, and the interarcuals should therefore be seen in a demonstration dissection.

The hypobranchial muscles are the common coracoarcuals, the coracomandibular, the coracohyoid, and the coracobranchials. They form a single complex which originates from the coracoid bar and inserts on the lower jaw and on the gill arches, performing various functions in connection with opening the mouth, swallowing, and expanding the gill pouches.

Remove the intermandibularis, the interhyoideus, and the superficial ventral constrictors to expose the gills, septa, and hypobranchial muscles as shown in the marginal illustration.

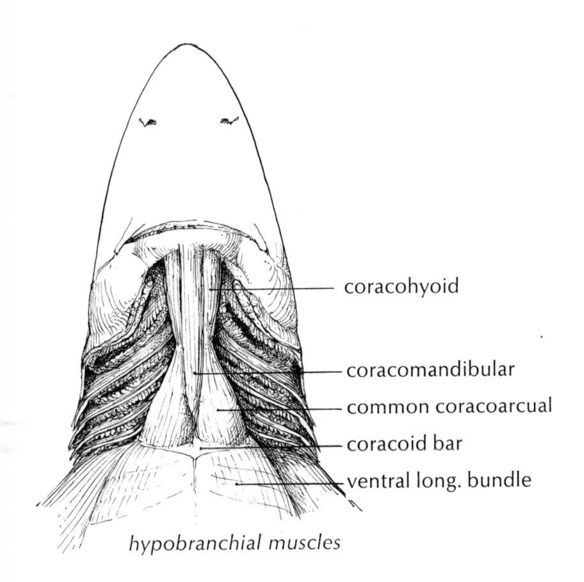


*hypobranchial muscles*

The common coracoarcuals lie immediately anterior to the coracoid bar, from which they take their origin. Anteriorly the fibers of the common coracoarcuals are continuous with those of the coracohyoid.

The coracomandibular is a slender muscle that lies in the midline. It originates deep to the intermandibularis from the fascia of the common coracoarcuals and inserts on the posterior edge of Meckel's cartilage.

The coracohyoids are paired muscles lying dorsal and lateral to the coracomandibular. They originate from the common coracoarcuals, with which their fibers are continuous, and insert on the basihyal, the ventral median cartilage of the hyoid arch.

The coracobranchials lie deep to the gill pouches and will be seen at a later stage in the dissection. See them as illustrated in Figure 10, page 17, and in Figure 17, page 29. The coracobranchials consist of five parts which insert on gill arches 2-5 and on the basibranchial cartilage. The afferent branchial arteries pass between them. The fifth coracobranchial and the common coracoarcual form the anterolateral wall of the pericardial cavity.

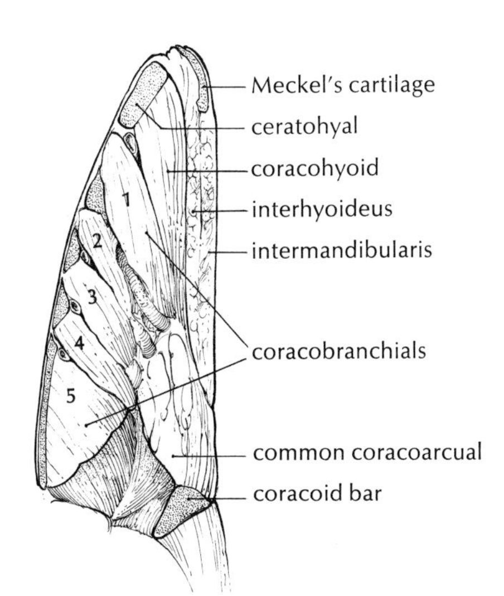


*sagittal section of hypobranchial muscles, cut left of midline*

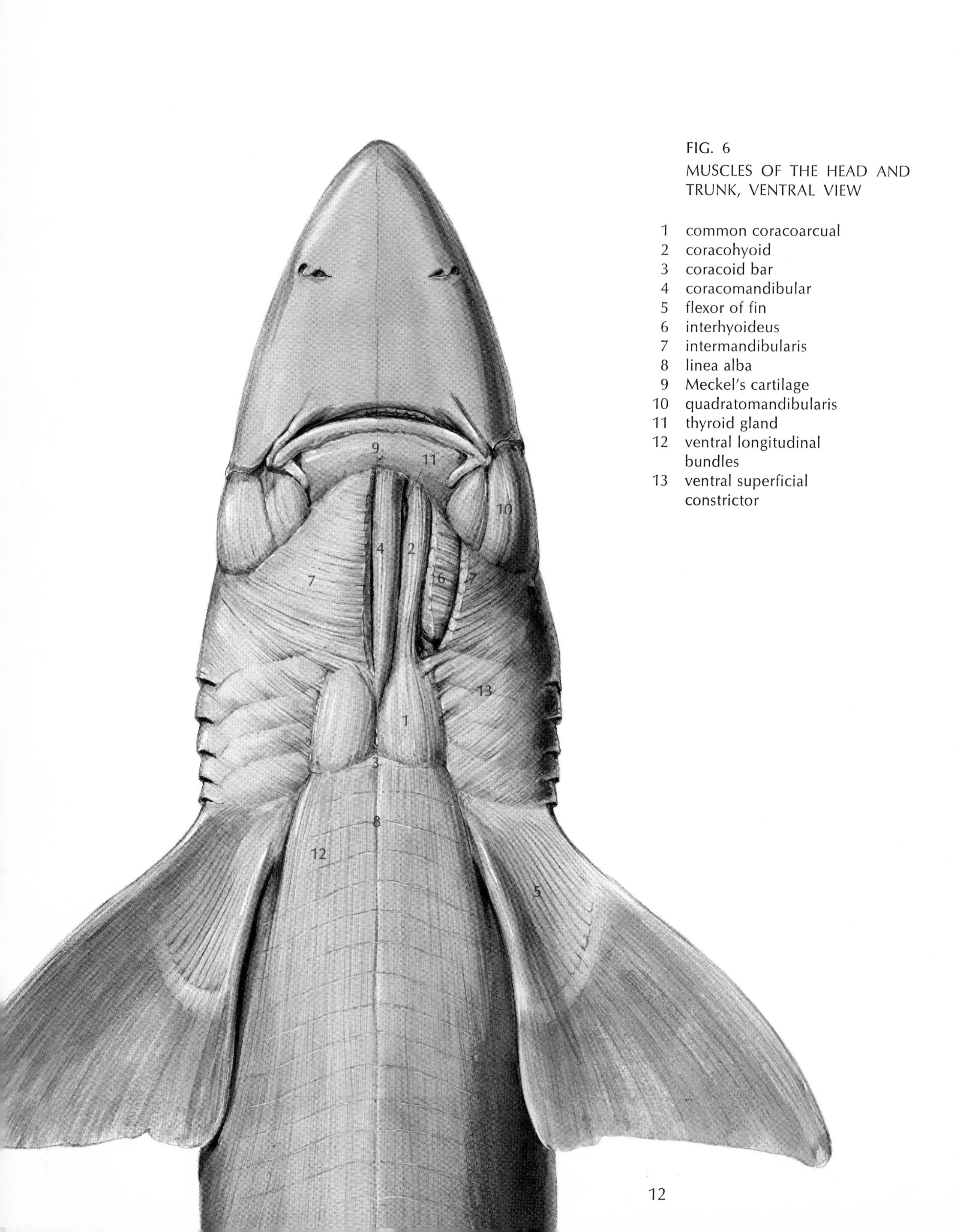


FIG. 6

MUSCLES OF THE HEAD AND TRUNK, VENTRAL VIEW

1 common coracoarcual
2 coracohyoid
3 coracoid bar
4 coracomandibular
5 flexor of fin
6 interhyoideus
7 intermandibularis
8 linea alba
9 Meckel's cartilage
10 quadratomandibularis
11 thyroid gland
12 ventral longitudinal bundles
13 ventral superficial constrictor

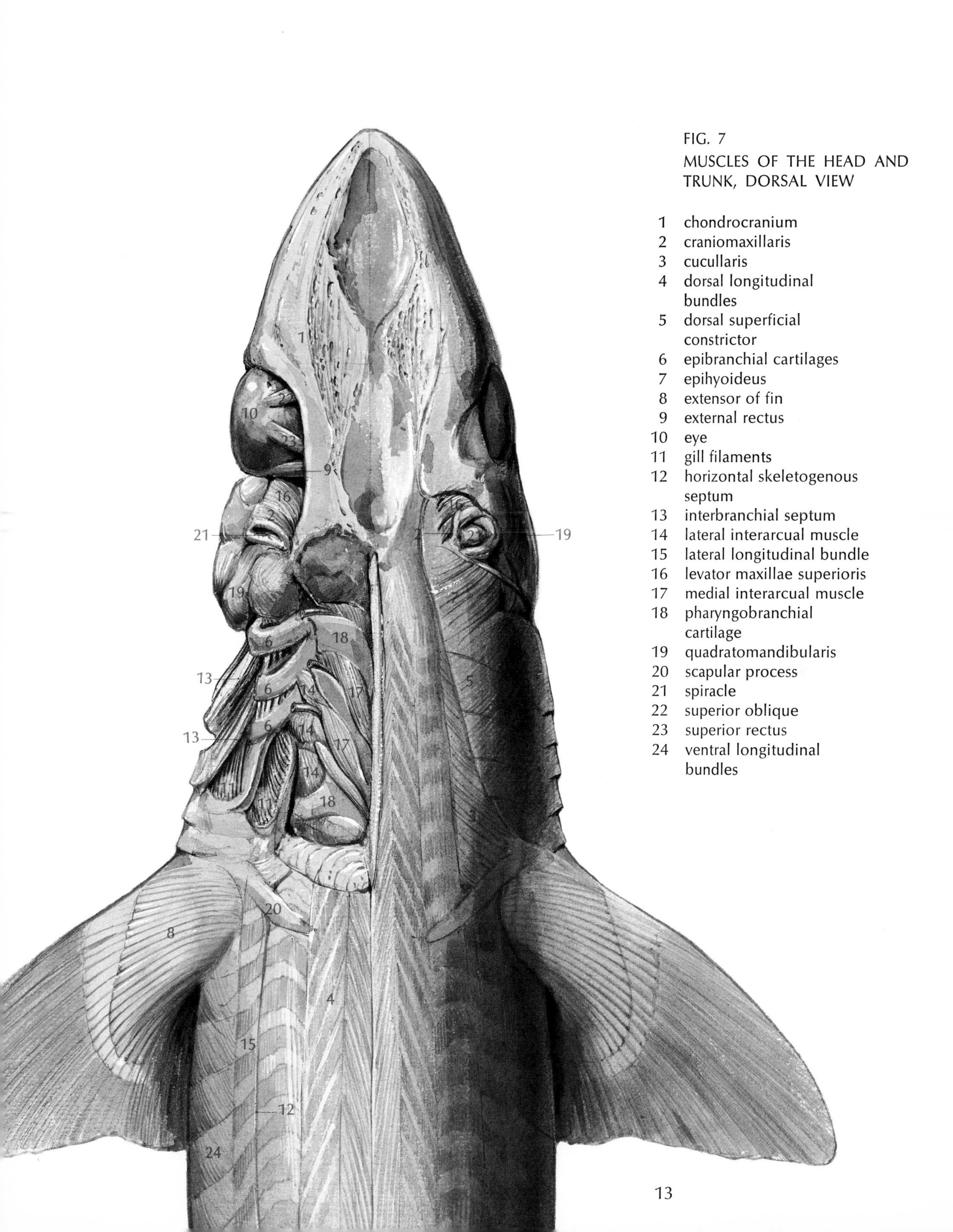


FIG. 7

MUSCLES OF THE HEAD AND TRUNK, DORSAL VIEW

1 chondrocranium
2 craniomaxillaris
3 cucullaris
4 dorsal longitudinal bundles
5 dorsal superficial constrictor
6 epibranchial cartilages
7 epihyoideus
8 extensor of fin
9 external rectus
10 eye
11 gill filaments
12 horizontal skeletogenous septum
13 interbranchial septum
14 lateral interarcual muscle
15 lateral longitudinal bundle
16 levator maxillae superioris
17 medial interarcual muscle
18 pharyngobranchial cartilage
19 quadratomandibularis
20 scapular process
21 spiracle
22 superior oblique
23 superior rectus
24 ventral longitudinal bundles

# THE DIGESTIVE AND RESPIRATORY SYSTEMS

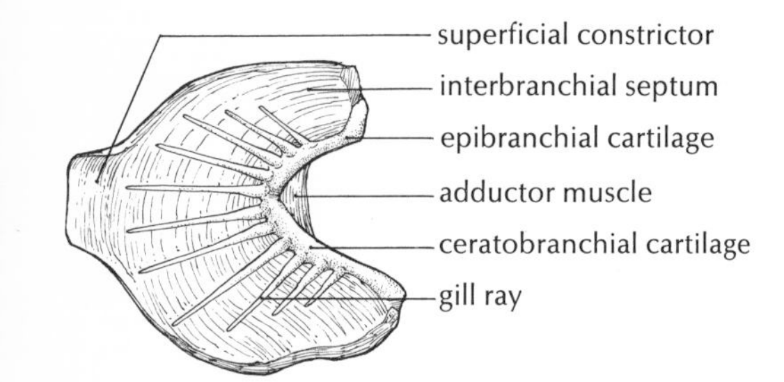


*interbranchial septum, posterior aspect, demibranch removed*

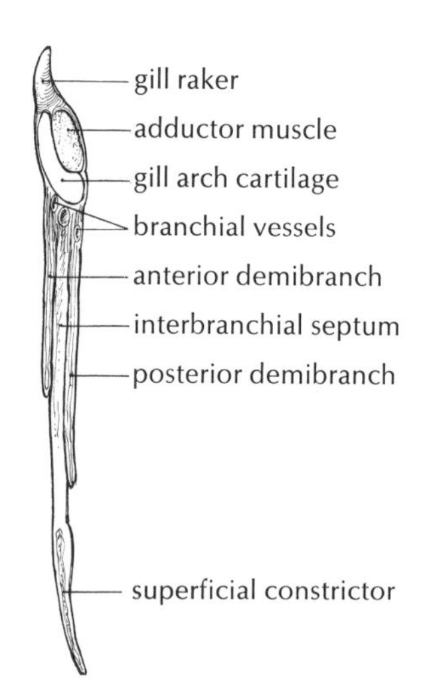


*section of branchial bar*

See demonstration dissections similar to Figures 8, 9, and 10, and identify the structures illustrated. The oral cavity of your specimen should be left intact at this time to preserve structures to be studied during the dissection of the circulatory system.

The oral cavity is the area enclosed by the jaws and hyoid arch. It is continuous posteriorly with the pharynx, which is the portion of the alimentary canal between the internal gill slits. Near the level of the pectoral girdle the pharynx joins the esophagus, which is distinguished by numerous small papillae. In the floor of the oral cavity is the practically immovable primary tongue. The hyoid arch lies within the anterior margin of the primary tongue. Palpate it and observe its cut surface.

In the roof of the oral cavity observe the paired internal openings of the spiracles. Posterior to them, in the lateral walls of the pharynx, are the five internal gill slits. The gill slits are supported by cartilaginous gill arches and guarded by small conical projections termed gill rakers. Each internal gill slit opens into a cavity termed a gill pouch; each gill pouch communicates with the outside via an external gill slit. The gill pouches are separated from each other by partitions termed branchial bars. Each branchial bar consists of a central sheet of connective tissue termed an interbranchial septum. Medially the septum is attached to a cartilaginous gill arch. Slender cartilaginous gill rays extend laterally from each gill arch, providing support for the interbranchial septum. On either side of each septum is a series of radial folds, termed gill lamellae, which contain the branchial vessels and capillaries. All the lamellae on both sides of a single interbranchial septum constitute a holobranch, or complete gill; the lamellae on either surface of the septum constitute a demibranch, or half gill.

Place an individual branchial bar in a dish of water and examine it under the dissecting microscope. It will be seen that the surface of each gill lamellae is covered with numerous minute secondary lamellae, which serve to increase the respiratory surface. Remove the demibranch from the posterior surface of the interbranchial septum and expose the gill rays as shown in the marginal illustration.

Examine the external opening of the spiracle and note that it is covered by a small spiracular valve, which bears a few rudimentary

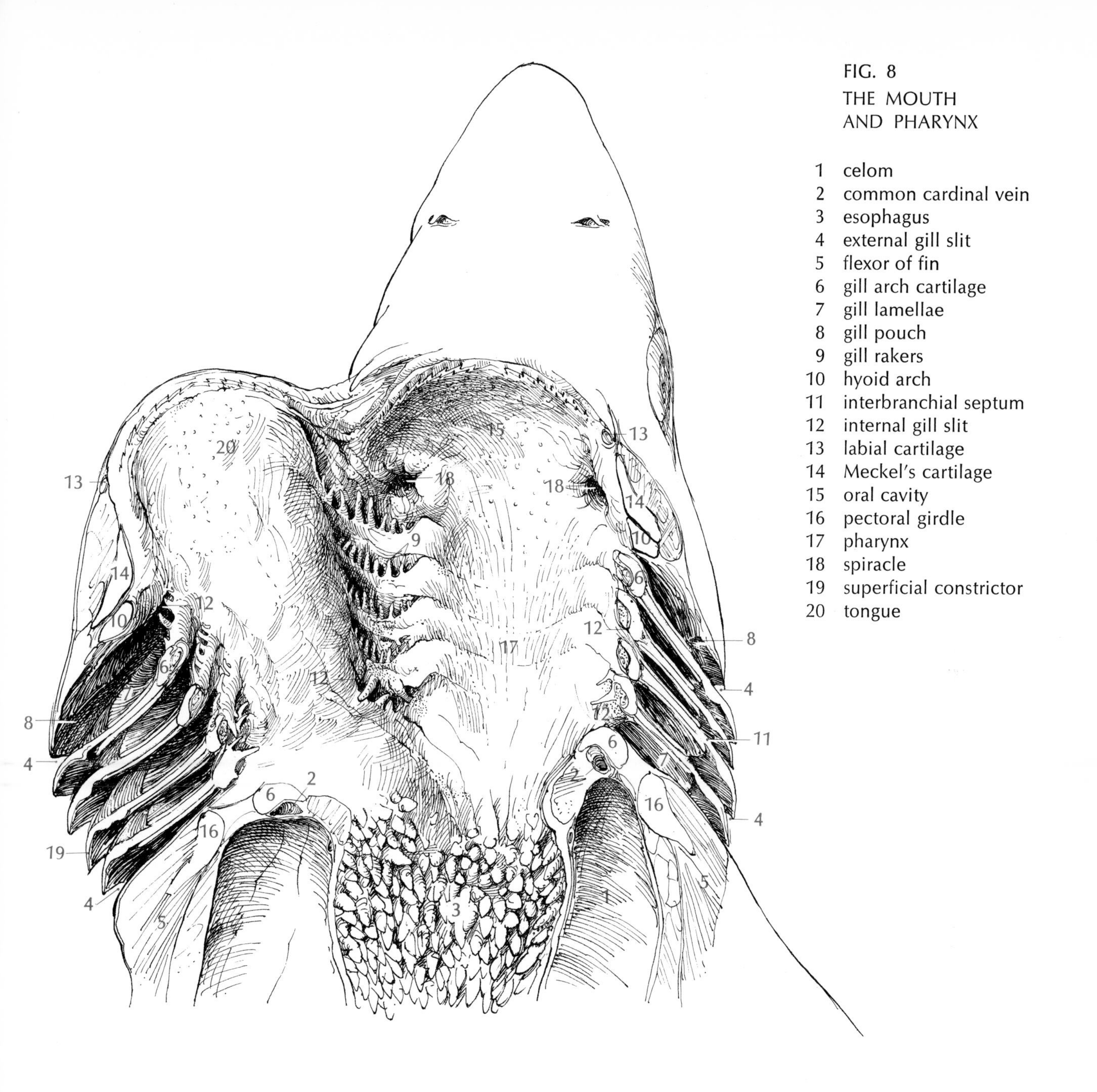

FIG. 8
THE MOUTH
AND PHARYNX

1 celom
2 common cardinal vein
3 esophagus
4 external gill slit
5 flexor of fin
6 gill arch cartilage
7 gill lamellae
8 gill pouch
9 gill rakers
10 hyoid arch
11 interbranchial septum
12 internal gill slit
13 labial cartilage
14 Meckel's cartilage
15 oral cavity
16 pectoral girdle
17 pharynx
18 spiracle
19 superficial constrictor
20 tongue

lamellae on its posterior surface. These lamellae are collectively termed a pseudobranch.

Observe a cross section of the branchial bar, as illustrated in the marginal diagram, and note the branchial vessels. The afferent branchial artery, which carries unoxygenated blood to the gill lamellae, lies in the middle of the septum. On either side of it are the efferent branchial arteries, which return oxygenated blood from the lamellae.

Starting just anterior to the cloacal opening, cut through the body wall and the puboischiac bar slightly to the right of the midline and extend the cut anteriorly through the coracoid bar and the common coracoarcual muscles. Observe and save the falciform ligament, which lies just posterior to the coracoid bar and extends from the ventral midline of the liver to the body wall. In the female

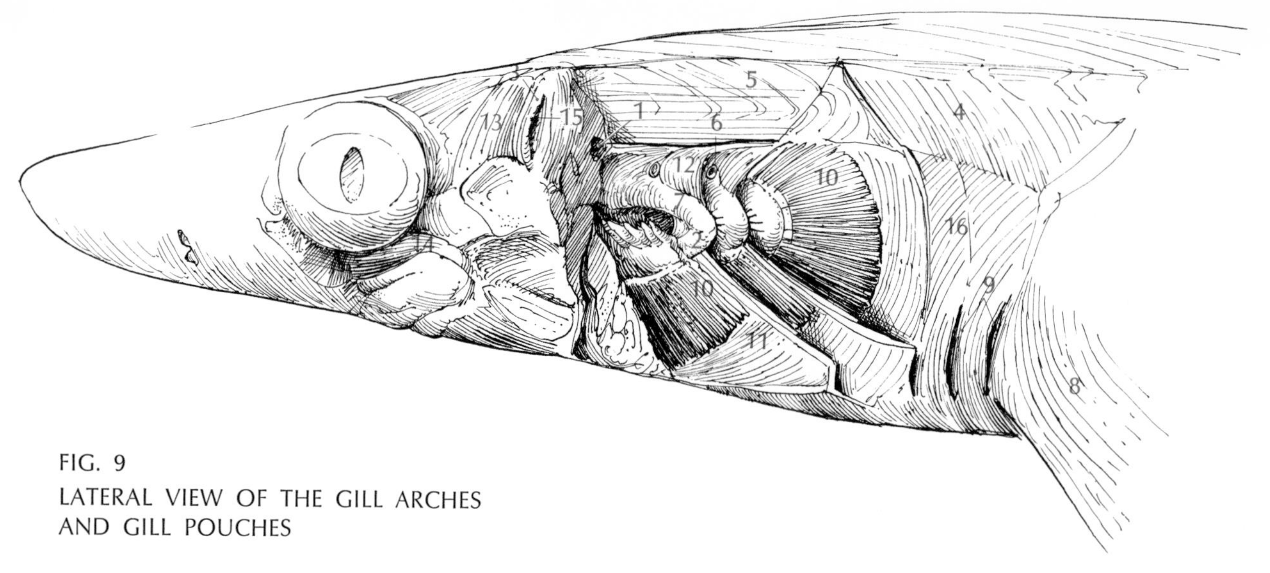

FIG. 9
LATERAL VIEW OF THE GILL ARCHES
AND GILL POUCHES

1 anterior cardinal sinus
2 ceratobranchial cartilage
3 craniomaxillaris
4 cucullaris
5 dorsal longitudinal bundle
6 efferent branchial artery
7 epibranchial cartilage
8 extensor of fin
9 external gill slits
10 gill lamellae (anterior demibranch)
11 interbranchial septum
12 lateral interarcual muscle
13 levator maxillae superioris
14 preorbitalis
15 spiracle
16 superficial constrictor

the oviducts curve around the anterior end of the liver, lying within the falciform ligament, and join to open by a common ostium, a funnel-shaped opening which lies in the posterior margin of the falciform ligament. If your specimen is a female probe the posterior margin of the falciform ligament and try to identify the ostium and the oviducts as illustrated in the marginal diagram. (They are difficult to identify in sexually immature animals.) Then trim the falciform ligament away from the ventral body wall, saving the oviducts, and pull back the body wall as illustrated in Figure 11 to expose the viscera.

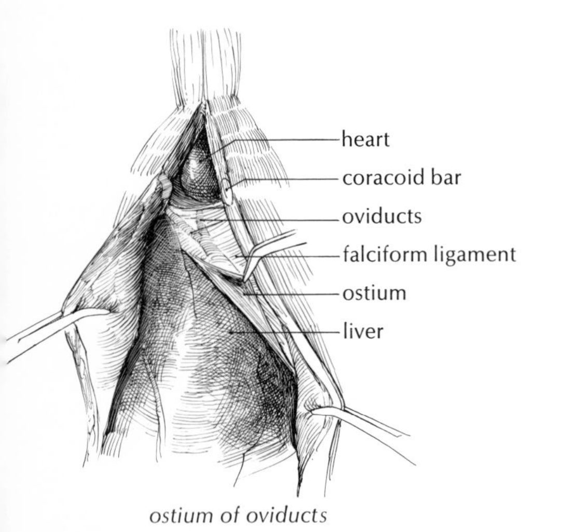


*ostium of oviducts*

The large cavity posterior to the coracoid bar is the pleuroperitoneal cavity. It contains the gonads, kidneys, liver, alimentary canal, and associated structures. The much smaller cavity anterior to the coracoid bar is the pericardial cavity. It contains the heart. The pleuroperitoneal cavity is separated from the pericardial cavity by a membranous partition termed the transverse septum.

Both the inside of the body wall and the viscera are lined by a serous membrane, the peritoneum. That portion of the peritoneum which lines the viscera is the visceral peritoneum, and that portion of the peritoneum which lines the abdominal wall is the parietal peritoneum. The peritoneal cavity is the potential space between the parietal and visceral layers. The peritoneal sheets that extend between the body wall and the viscera are termed mesenteries, ligaments, and omenta. Within these sheets are the vessels, nerves, and lymphatics which supply the viscera.

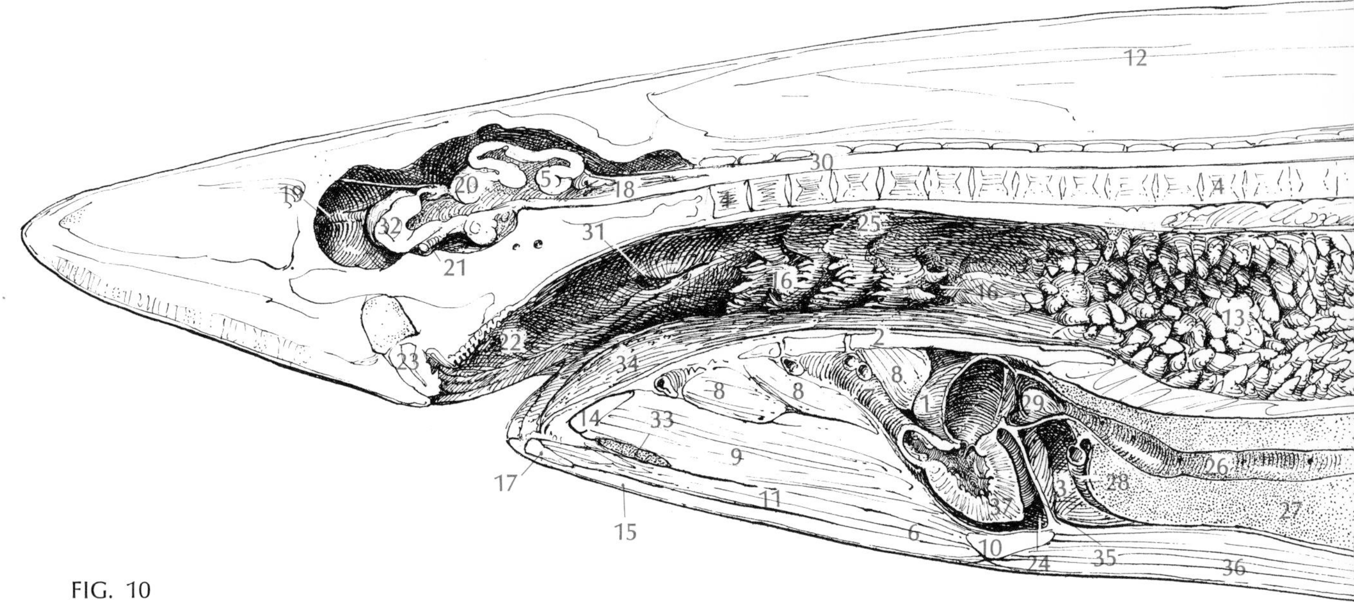


FIG. 10
SAGITTAL SECTION OF THE HEAD

1 atrium
2 basibranchial cartilage
3 celom
4 centrum
5 cerebellum
6 common coracoarcual
7 conus arteriosus
8 coracobranchial
9 coracohyoid
10 coracoid bar
11 coracomandibular
12 dorsal longitudinal bundles
13 esophagus
14 hyoid arch
15 intermandibularis
16 internal gill slit
17 Meckel's cartilage
18 medulla
19 olfactory tract
20 optic lobe
21 optic nerve
22 oral cavity
23 palatoquadrate cartilage
24 pericardial cavity
25 pharynx
26 right hepatic sinus
27 right lobe of liver
28 right oviduct
29 sinus venosus
30 spinal cord
31 spiracle
32 telencephalon
33 thyroid gland
34 tongue
35 transverse septum
36 ventral longitudinal bundle
37 ventricle

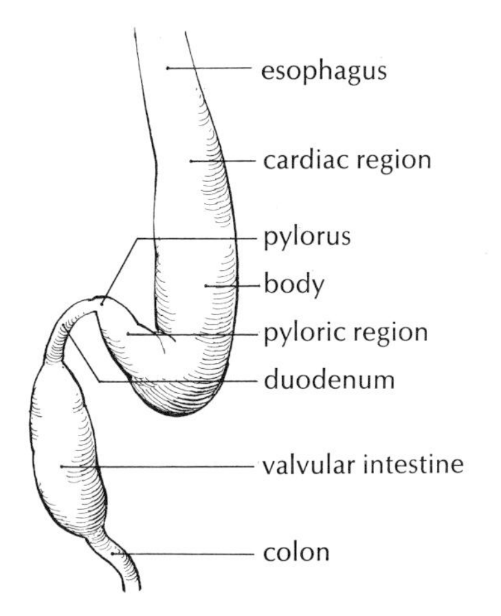


*alimentary canal*

The liver is attached to the transverse septum by the coronary ligament and to the ventral body wall by the falciform ligament. Its prominent right and left lobes extend posteriorly throughout most of the pleuroperitoneal cavity. Between the right and left lobes, near the ventral midline, is the median lobe. The gallbladder is a thin tubular sac that lies along the right margin of the median lobe.

The short esophagus connects the oral cavity and pharynx with the stomach, which lies between the right and left lobes of the liver. No definite line of transition can be observed externally between the esophagus and the stomach. The anterior portion of the stomach is termed the cardiac region; posterior to this is the body of the stomach. Near the posterior end of the left lobe of the liver the stomach bends abruptly forward and diminishes in diameter. This part is termed the pyloric region. The left-hand margin of the stomach is termed the greater curvature; the right-hand margin is termed the lesser curvature. The stomach terminates at the pylorus, a sphincter muscle which regulates the flow of food from the stomach to the intestine.

The intestine is the part of the alimentary canal lying distal to the pylorus; it consists of the duodenum, the valvular intestine, and the colon. The duodenum is a short, curved section, most of which is concealed by the ventral lobe of the pancreas. Bile is conveyed to the duodenum from the gallbladder by the bile duct, which lies along the hepatic portal vein. Examine the point at which the bile duct enters the dorsal wall of the duodenum.

The surface of the valvular intestine is marked by rings which correspond to the attachment of the inner folds of the spiral valve. At the posterior end of the pleuroperitoneal cavity the valvular intestine merges with the narrower colon. The colon and the urogenital ducts open into a short chamber, the cloaca, which communicates with the outside via the cloacal opening between the pelvic fins. A small rectal gland lies near the colon and is attached to it by a duct.

The dark-colored organ near the posterior end of the stomach is the spleen, part of the lymphatic system.

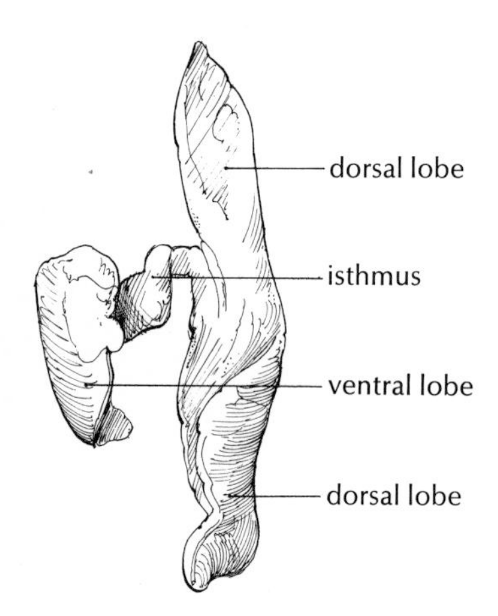

*pancreas, ventral view*

The pancreas consists of a ventral lobe, which lies ventral to the duodenum, and an elongated dorsal lobe, which lies dorsal to the duodenum and extends posteriorly to the spleen. The dorsal and ventral lobes are connected by a slender band of pancreatic tissue termed the isthmus. Pancreatic secretions enter the duodenum through a small pancreatic duct which lies embedded in the pancreas near the posterior margin of the ventral lobe.

Near the posterolateral margin of the cloacal opening are two small passages termed the abdominal pores. Pass a blunt probe through one of them and confirm the fact that it forms a connection between the pleuroperitoneal cavity and the exterior.

Lift the lobes of the liver and identify the gonads (testes or ovaries), which will be seen against the dorsal body wall at the anterior end of the pleuroperitoneal cavity. The kidneys are dark, elongated structures lying on either side of the dorsal midline and covered by peritoneum. In mature animals the oviduct (female) or archinephric duct (male) may be seen extending from the gonad to the cloaca (see Fig. 19, p. 33).

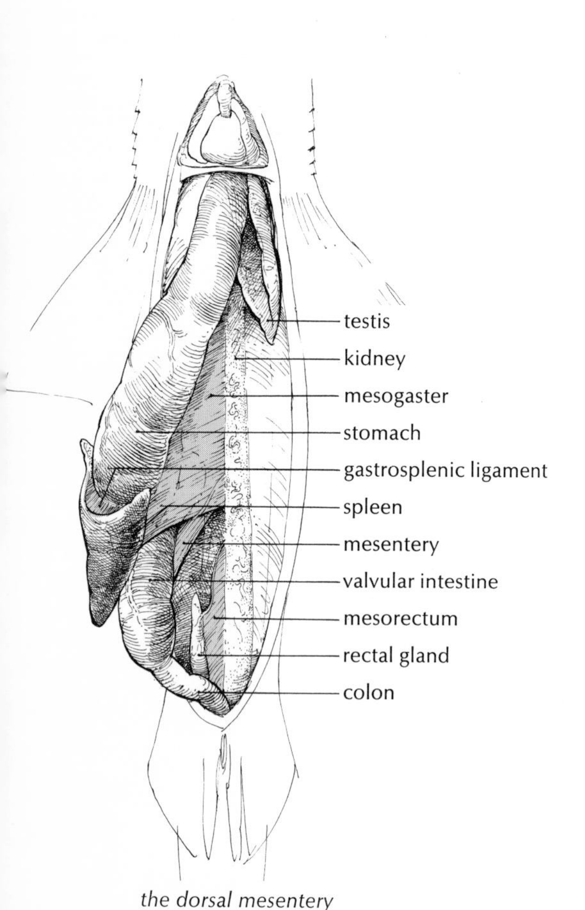

*the dorsal mesentery*

The dorsal mesentery extends from the mid-dorsal line of the body wall to the alimentary canal. Its derivatives are the mesogaster, the gastrosplenic ligament, the mesentery proper, and the mesorectum. Pull the stomach and liver to the right and examine the portion of the dorsal mesentery which supports the stomach. This is the mesogaster, and the part of the mesogaster extending directly between the spleen and the stomach is the gastrosplenic ligament. Near the posterior end of the mesogaster two arteries leave the dorsal body wall and extend to the alimentary canal; they are the anterior mesenteric artery, which supplies the valvular intestine, and the lienogastric artery, which supplies the spleen and the posterior end of the stomach.

Pull the stomach and intestine to the left and examine the portion of the dorsal mesentery which supports the valvular intestine and duodenum. It is the mesentery in the limited sense. (The term "mesentery" is used in two ways: in the limited sense it means the dorsal mesentery of the intestine, but more generally it refers to the membranous attachment of any of the abdominal viscera.) Observe that the dorsal lobe of the pancreas is supported by the mesogaster,

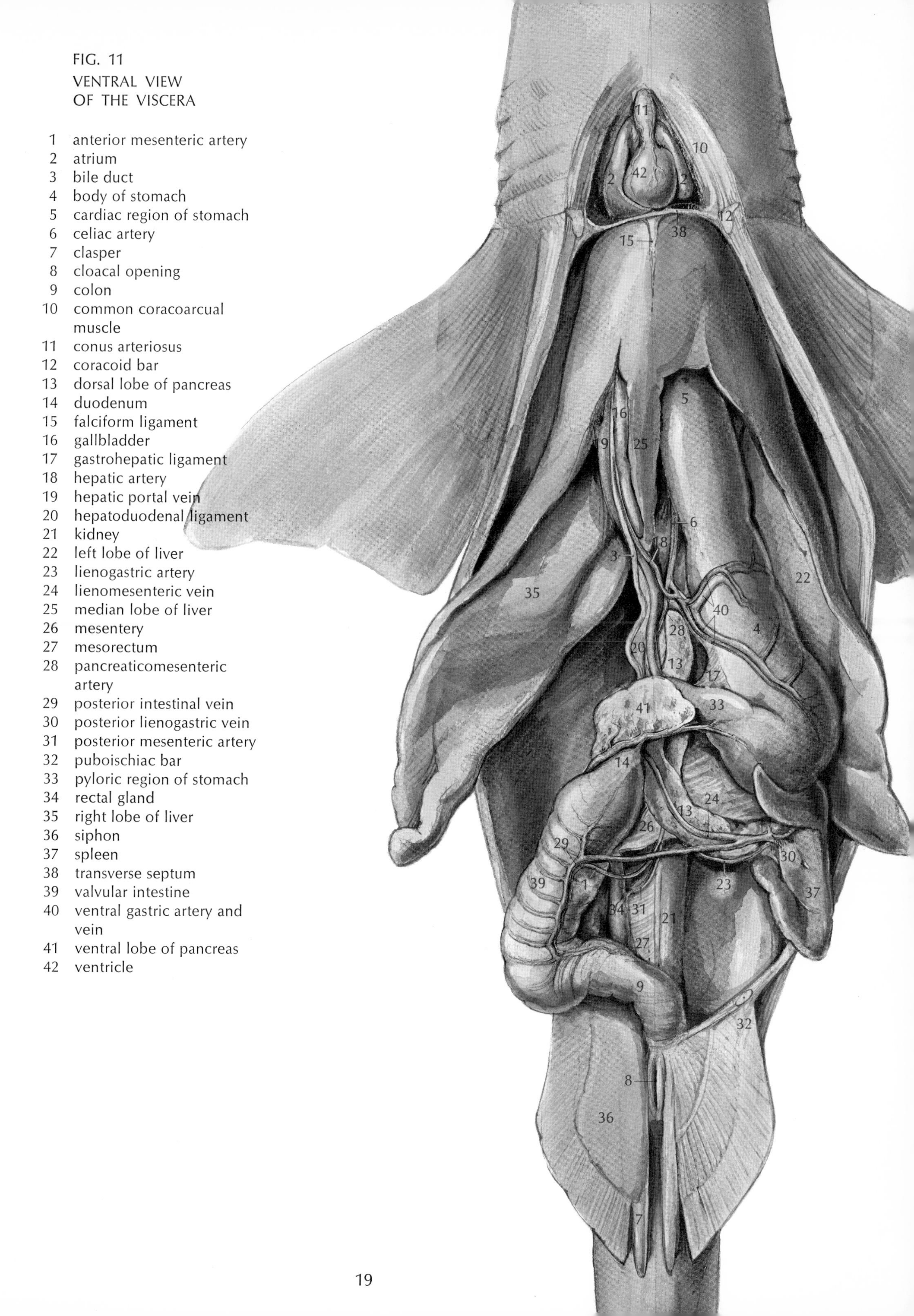

FIG. 11
VENTRAL VIEW
OF THE VISCERA

1 anterior mesenteric artery
2 atrium
3 bile duct
4 body of stomach
5 cardiac region of stomach
6 celiac artery
7 clasper
8 cloacal opening
9 colon
10 common coracoarcual muscle
11 conus arteriosus
12 coracoid bar
13 dorsal lobe of pancreas
14 duodenum
15 falciform ligament
16 gallbladder
17 gastrohepatic ligament
18 hepatic artery
19 hepatic portal vein
20 hepatoduodenal ligament
21 kidney
22 left lobe of liver
23 lienogastric artery
24 lienomesenteric vein
25 median lobe of liver
26 mesentery
27 mesorectum
28 pancreaticomesenteric artery
29 posterior intestinal vein
30 posterior lienogastric vein
31 posterior mesenteric artery
32 puboischiac bar
33 pyloric region of stomach
34 rectal gland
35 right lobe of liver
36 siphon
37 spleen
38 transverse septum
39 valvular intestine
40 ventral gastric artery and vein
41 ventral lobe of pancreas
42 ventricle

which fuses with the mesentery near this point, forming a shallow pouch dorsal to the stomach. The dorsal mesentery is incomplete at the posterior end of the alimentary canal so that there is no direct mesenteric support for the posterior end of the valvular intestine. The rectal gland and the posterior part of the colon are supported by a derivative of the dorsal mesentery, the mesorectum. The posterior mesenteric artery runs along the anterior margin of the mesorectum.

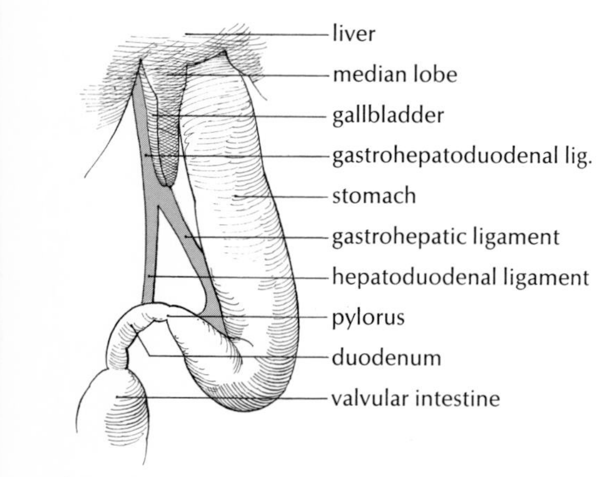


*gastrohepatoduodenal ligament*

The derivatives of the ventral mesentery are the falciform ligament of the liver and the gastrohepatoduodenal ligament. The falciform ligament lies near the anterior end of the liver in the ventral midline and attaches the liver to the ventral body wall. The gastrohepatoduodenal ligament (or lesser omentum) is a slender membrane which extends from the liver to the stomach and duodenum. It is shaped like an inverted letter Y. Anteriorly it is a single structure which supports the bile duct, the hepatic artery, and the hepatic portal vein. Posteriorly it divides into two parts: the hepatoduodenal ligament and the gastrohepatic ligament. The hepatoduodenal ligament is attached to the duodenum and follows the bile duct and the hepatic portal vein; the gastrohepatic ligament is attached to the stomach and extends across the lesser curvature from the body to the pyloric region. Both the hepatoduodenal ligament and the gastrohepatic ligament support branches of the celiac artery.

The gonads are supported by mesenteries which arise from the dorsal body wall near the mesogaster. The mesentery of the testis is the mesorchium; that of the ovary is the mesovarium. In mature females each oviduct is supported by a mesentery termed the mesotubarium.

Refer to Figure 12. The celiac artery, lienomesenteric artery, anterior and posterior mesenteric arteries, and the hepatic portal vein will be described at this time because of their close association with the alimentary canal.

The celiac artery, a branch of the dorsal aorta, emerges between the stomach and the middle lobe of the liver. It passes posteriorly, lying along the lesser curvature of the stomach, to the anterior end of the pancreas. At this point, near the origin of the hepatic portal vein, it divides into three branches: the gastric, the hepatic, and the pancreaticomesenteric.

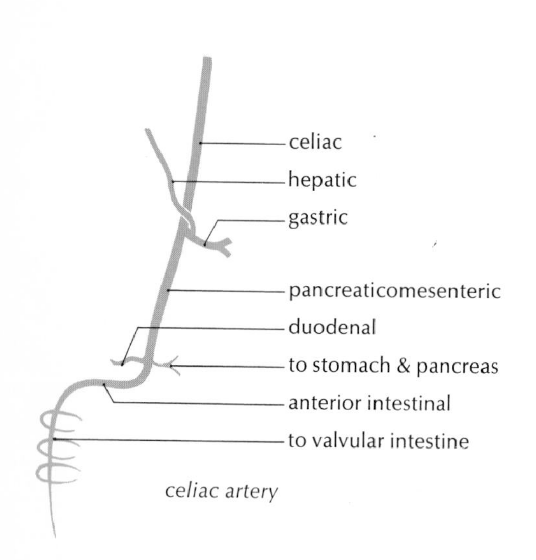


*celiac artery*

The gastric artery supplies the stomach; it divides into dorsal and ventral branches which lie on the dorsal and ventral sides of the stomach.

The hepatic artery turns anteriorly and runs parallel to the bile duct and the hepatic portal vein to supply the liver. The pancreaticomesenteric artery passes posteriorly, dorsal to the pylorus, lying next to the pancreaticomesenteric vein. It gives a duodenal artery to the duodenum and small branches to the pyloric region of the stomach and to the pancreas. It then continues as the anterior intestinal artery, lying next to the anterior intestinal vein, and hidden by the posterior margin of the ventral lobe of the pancreas. It supplies the valvular intestine, to which it gives branches that lie along the lines of attachment of the spiral valve.

Near the dorsal edge of the mesogaster the dorsal aorta gives off two branches: the anterior mesenteric and the lienogastric arteries. The anterior mesenteric artery supplies the left side of the valvular intestine, lying along the posterior intestinal vein. The

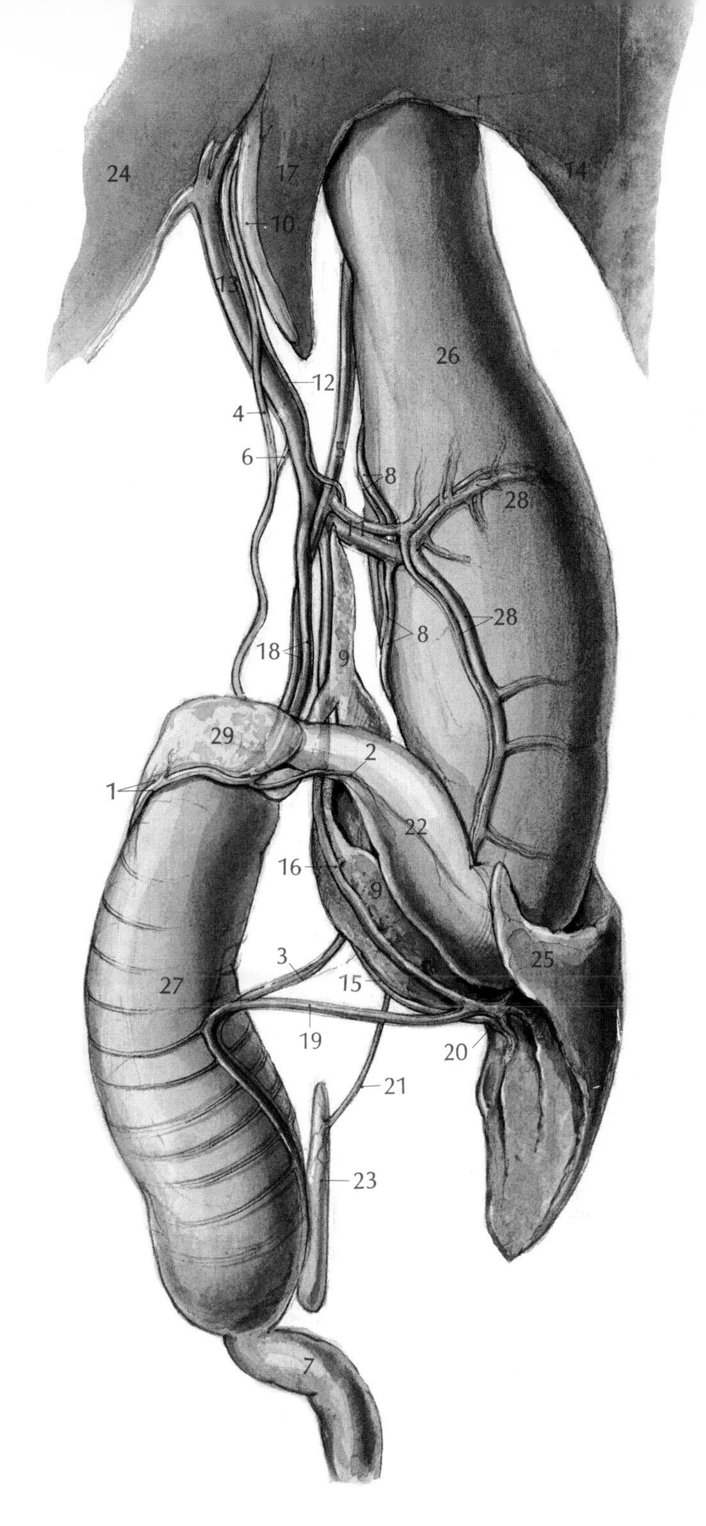


FIG. 12
BLOOD SUPPLY OF
THE ALIMENTARY CANAL

1 anterior intestinal artery and vein
2 anterior lienogastric vein
3 anterior mesenteric artery
4 bile duct
5 celiac artery
6 choledochal vein
7 colon
8 dorsal gastric artery and vein
9 dorsal lobe of pancreas
10 gallbladder
11 gastric artery and vein
12 hepatic artery
13 hepatic portal vein
14 left lobe of liver
15 lienogastric artery
16 lienomesenteric vein
17 median lobe of liver
18 pancreaticomesenteric artery and vein
19 posterior intestinal vein
20 posterior lienogastric vein
21 posterior mesenteric artery
22 pyloric region of stomach
23 rectal gland
24 right lobe of liver
25 spleen
26 stomach
27 valvular intestine
28 ventral gastric artery and vein
29 ventral lobe of pancreas

lienogastric artery lies along the posterior margin of the dorsal lobe of the pancreas; it supplies the stomach and spleen. Posterior to the anterior mesenteric and lienogastric arteries the dorsal aorta gives off a small posterior mesenteric artery which runs along the anterior margin of the mesorectum to supply the rectal gland.

The hepatic portal vein conveys blood from the alimentary canal, spleen, pancreas, and rectal gland to the liver, within which it ramifies to form a system of capillaries and blood sinuses. It is formed by the union of three veins: the gastric vein, the lienomesenteric vein, and the pancreaticomesenteric vein.

Trace the gastric vein to the stomach, noting that it passes dorsal to the celiac artery. It is formed near the middle of the stomach by the union of two dorsal gastric veins from the dorsal side of the stomach, and two ventral gastric veins from the ventral side of the stomach.

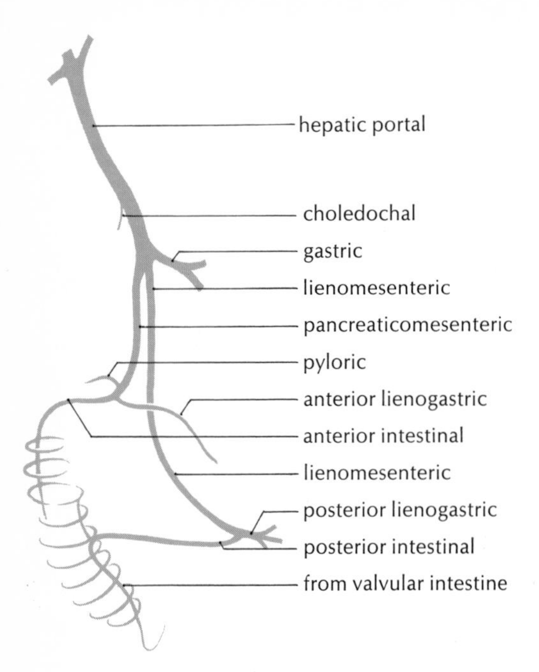


*hepatic portal vein*

The lienomesenteric vein lies on the surface of the dorsal lobe of the pancreas, from which it receives numerous small pancreatic veins. It passes dorsal to the duodenum and then turns to the left and extends to the spleen. Near the spleen it is formed by the union of two veins: the posterior intestinal vein and the posterior lienogastric vein. The posterior intestinal vein lies along the left side of the valvular intestine, from which it receives numerous small transverse tributaries originating along the lines of attachment of the spiral valve. The posterior lienogastric vein originates from the union of several short veins which return blood from the spleen and the posterior end of the stomach.

The pancreaticomesenteric vein passes dorsal to the pylorus, lying along the isthmus of the pancreas. It receives the pyloric vein from the pylorus and duodenum, and the intraintestinal vein from the interior of the spiral valve. It is formed just posterior to the pylorus by the union of the anterior lienogastric and anterior intestinal veins. The anterior intestinal vein originates from the right side of the valvular intestine and lies along the posterior margin of the ventral lobe of the pancreas, from which it receives several tributaries. The anterior lienogastric vein originates from the spleen and lies along the ventral surface of the pyloric region of the stomach.

Just anterior to the union of the gastric, lienomesenteric, and pancreaticomesenteric veins, the hepatic portal vein receives a small choledochal vein from the posterior end of the bile duct.

Remove the alimentary canal, being careful to leave the cloaca and the liver intact. Cut open the alimentary canal as illustrated in the marginal diagram. The mucous membrane of the esophagus bears small projections termed papillae; in the stomach, the mucous membrane forms folds termed rugae. Within the valvular intestine is the spiral valve, formed by overlapping folds which serve to increase the absorptive surface.

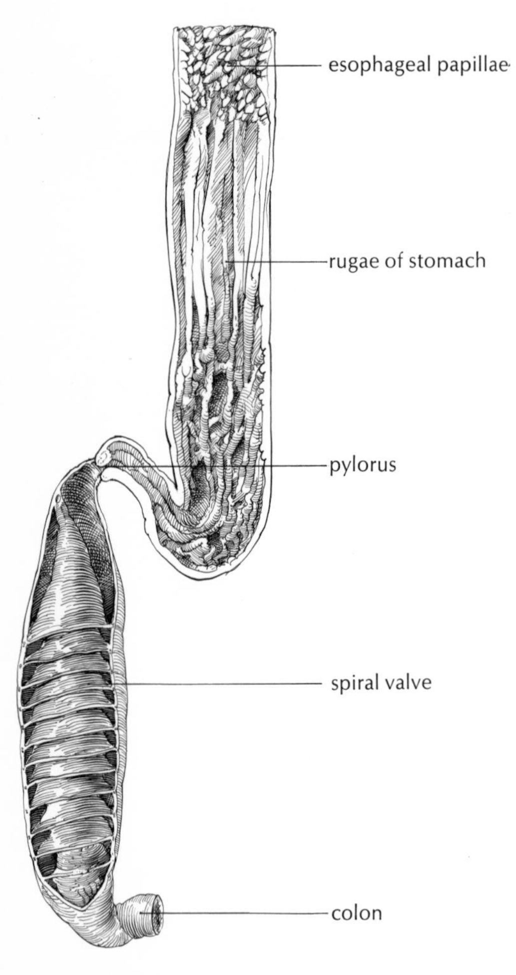


*stomach and valvular intestine*

# THE CIRCULATORY SYSTEM

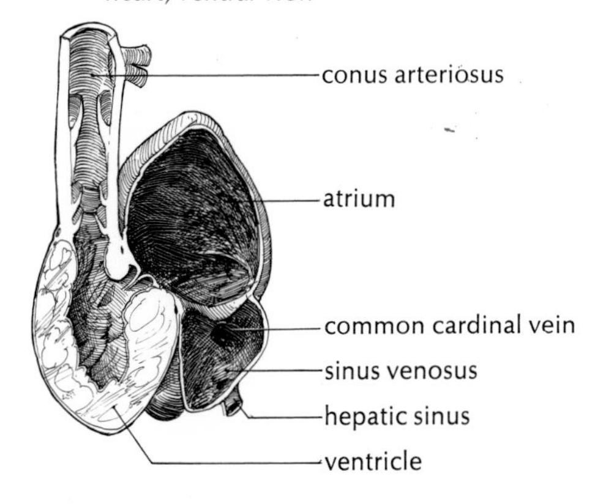


*heart, ventral view*



*heart, sagittal section*

The pericardial cavity is the division of the embryonic celom which lies anterior to the transverse septum and contains the heart. The dorsolateral walls of the pericardial cavity are formed by the basibranchial cartilage and the fifth coracobranchial muscles. The ventrolateral walls are formed by the common coracoarcuals, and the posterior wall is formed by the transverse septum.

Both the heart and the walls of the pericardial cavity are lined by a membrane termed the pericardium. The layer of pericardium that lines the walls of the pericardial cavity is the parietal pericardium, and the layer that lines the heart is the visceral pericardium. The parietal pericardium can be dissected away from the walls of the pericardial cavity, but the visceral pericardium is inseparably fused with the walls of the heart. The parietal and visceral layers are continuous with each other at the anterior and posterior ends of the heart, where the conus arteriosus and sinus venosus are attached to the walls of the pericardial cavity.

Venous blood returning to the heart first enters the sinus venosus, a thin-walled triangular sac dorsal to the ventricle. To see the sinus venosus, lift the ventricle and pull it to one side as illustrated in Figure 15.

From the sinus venosus blood passes to the atrium, a large, dark-colored chamber which lies dorsal to the ventricle and extends laterally on either side of it. Observe that the walls of the atrium are very thin, whereas the walls of the ventricle are thick and muscular. Blood passes from the atrium into the ventricle, and is pumped by the ventricle into the conus arteriosus. Five pairs of afferent branchial arteries convey unoxygenated blood from the conus arteriosus to the gill filaments, where the exchange of respiratory gases occurs.

The veins of the dogfish are for the most part sinuses or cavities without clearly defined walls. Unless they are injected they are difficult to trace, and for this reason some specimens with injected veins should be available for reference. If the veins in your specimen are not injected, trace them as far as possible by opening them with small scissors and following the cavities with a probe.

Refer to Figure 13. Two pairs of veins enter the sinus venosus: the

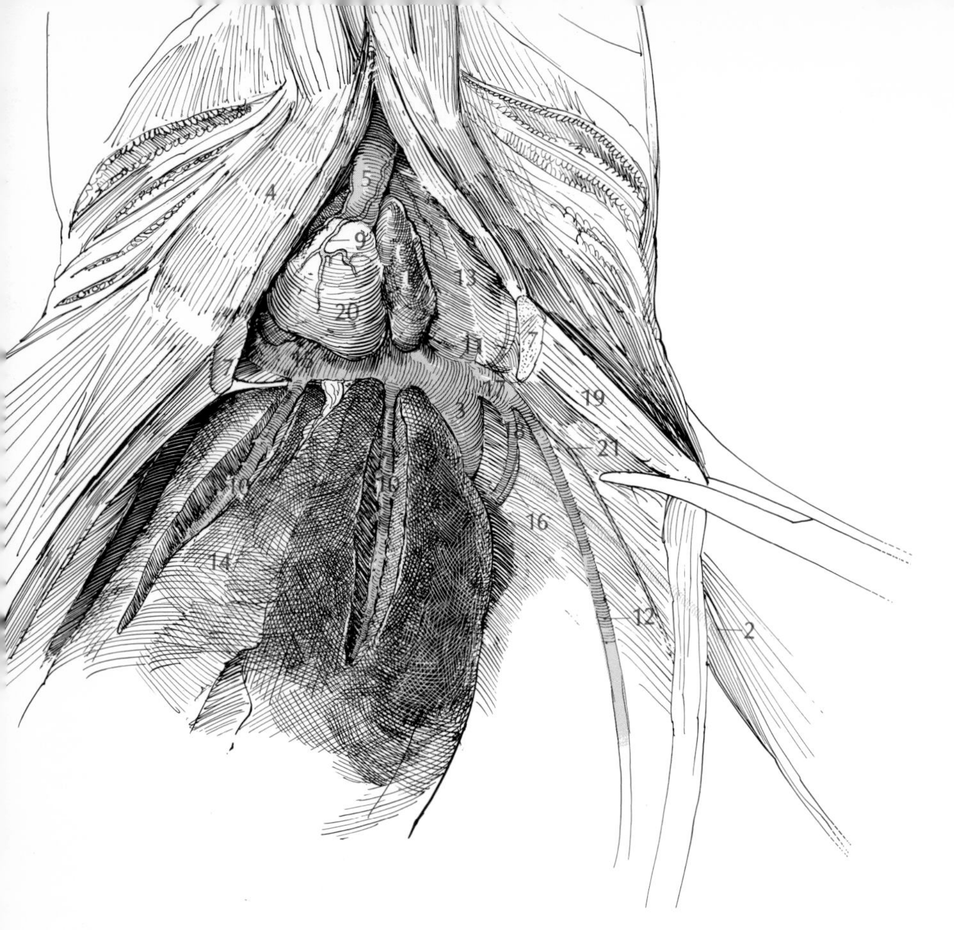

FIG. 13
THE HEART AND VENTRAL VEINS

1 atrium
2 brachial vein
3 common cardinal vein
4 common coracoarcual muscle
5 conus arteriosus
6 coracohyoid muscle
7 coracoid bar
8 coracomandibular muscle
9 coronary artery
10 hepatic sinus
11 inferior jugular vein
12 lateral abdominal vein
13 lateral wall of pericardial cavity
14 liver
15 sinus venosus
16 subclavian artery
17 subclavian vein
18 subscapular vein
19 ventral longitudinal bundles
20 ventricle
21 ventrolateral artery

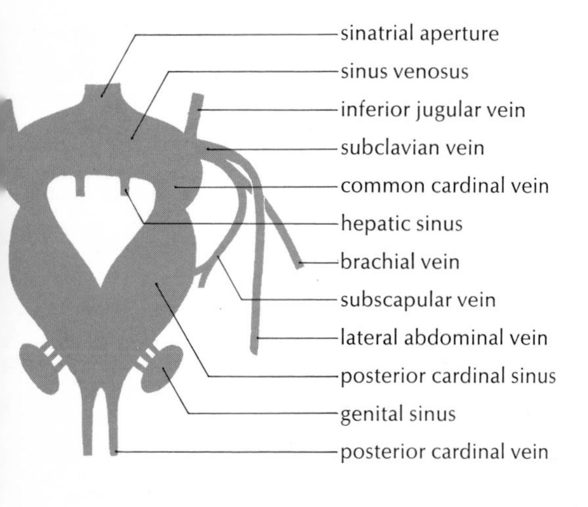


*sinus venosus and posterior cardinal sinuses*

hepatic veins and the common cardinal veins (ducts of Cuvier).

Probe between the sinus venosus and the liver to locate the points where the hepatic sinuses enter the sinus venosus. Trace them into the liver by probing and slitting open the ventral surface of the liver. Venous blood from the viscera of the pleuroperitoneal cavity is returned to the liver by the hepatic portal vein; within the liver it passes through a system of sinuses and capillaries, and then returns to the heart via the hepatic sinuses.

The common cardinal veins return blood from the body. They are broad, short vessels which curve dorsally around the body wall and fuse with the large posterior cardinal sinuses, to be described below.

Slit open the ventral side of the sinus venosus and identify the opening of the inferior jugular vein, which enters the anterior aspect of the common cardinal vein at the point where the common cardinal vein joins the sinus venosus. The inferior jugular vein extends anteriorly, dorsal to the muscles forming the lateral wall of the pericardial cavity, and receives tributaries from the floor of the mouth, the pharynx, and the branchial region.

Just lateral to the entrance of the inferior jugular vein the common cardinal vein receives the short subclavian vein which is formed by the union of three veins: the brachial vein, the subscapular vein, and the lateral abdominal vein.

The lateral abdominal vein originates near the pelvic girdle and passes anteriorly along the body wall, receiving small parietal branches from the myotomes.

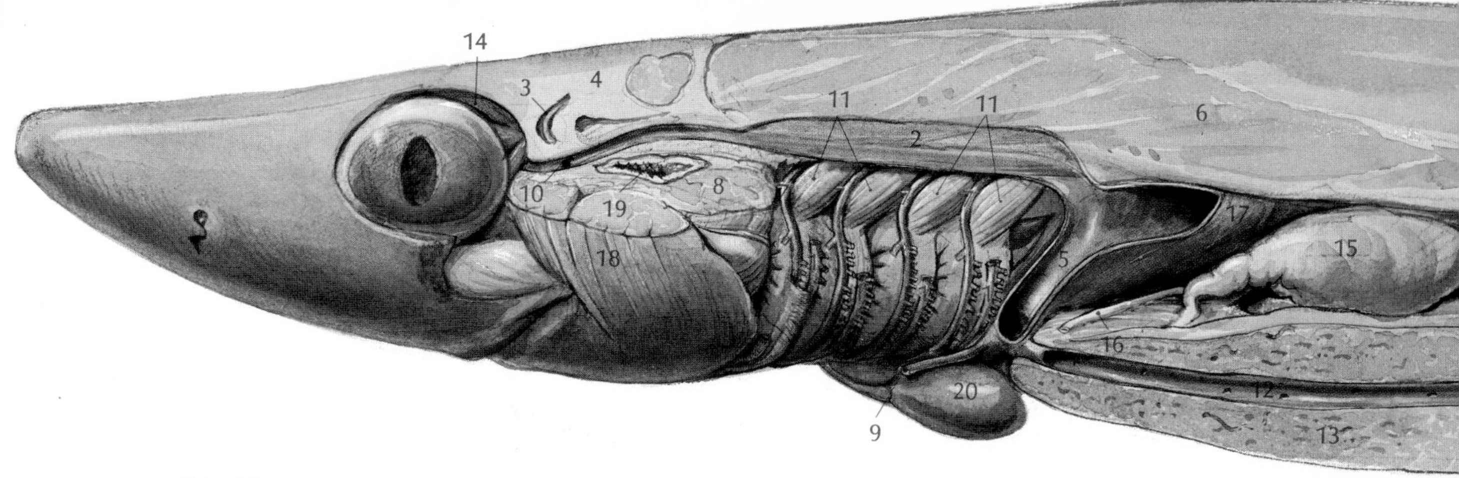

FIG. 14
THE ANTERIOR CARDINAL SINUS

1 afferent branchial artery
2 anterior cardinal sinus
3 anterior semicircular duct
4 chondrocranium
5 common cardinal vein
6 dorsal longitudinal bundles
7 efferent branchial artery
8 epihyoidean muscle
9 inferior jugular vein
10 interorbital sinus opening into ant. cardinal sinus
11 lateral interarcual muscle
12 left hepatic sinus
13 left lobe of liver
14 orbital sinus
15 ovary
16 oviduct
17 posterior cardinal sinus
18 quadratomandibularis muscle
19 spiracle
20 ventricle

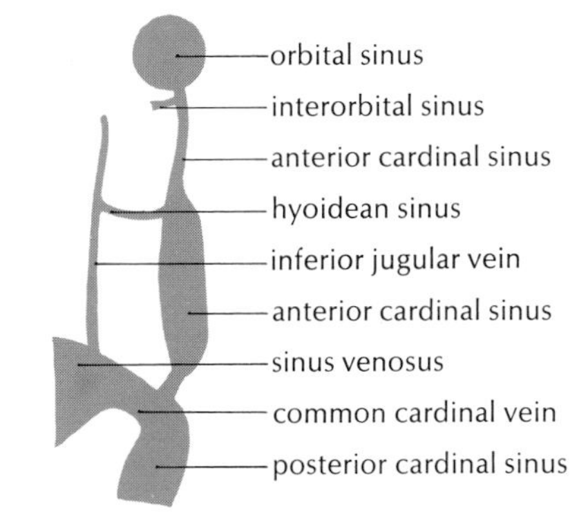


*anterior cardinal sinus, lateral view*

The subscapular vein originates near the dorsal line, where it communicates with the posterior cardinal sinus, and lies along the posterior margin of the coracoid bar, passing ventrally to join the subclavian vein.

The brachial vein lies along the posterior side of the pectoral fin, passing medially to join the lateral abdominal vein. In many individuals it joins the subscapular vein.

Refer to Figure 14. Make a shallow cut along the ventral margin of the cucullaris muscle. Your knife will enter a space, the anterior cardinal sinus, which lies dorsal to the pharyngobranchial cartilages. The anterior cardinal sinus receives tributaries which return venous blood from the brain, eye, and head. Posteriorly it communicates with the common cardinal vein via a narrow channel. Identify this communication by probing. Expose the anterior cardinal sinus by removing the gill lamellae and the interbranchial septa, and by cutting the jaw muscles and the chondrocranium.

Anteriorly the anterior cardinal sinus extends medial to the spiracle and communicates with the orbital sinus, which surrounds the eyeball and its muscles. The right and left anterior cardinal sinuses are connected to each other by the interorbital sinus, which opens into the ventral aspect of the anterior cardinal sinus just posterior to the orbit. Each anterior cardinal sinus is also connected to the inferior jugular vein by a hyoidean sinus, which opens into the ventral aspect of the anterior cardinal sinus between the second and third gill

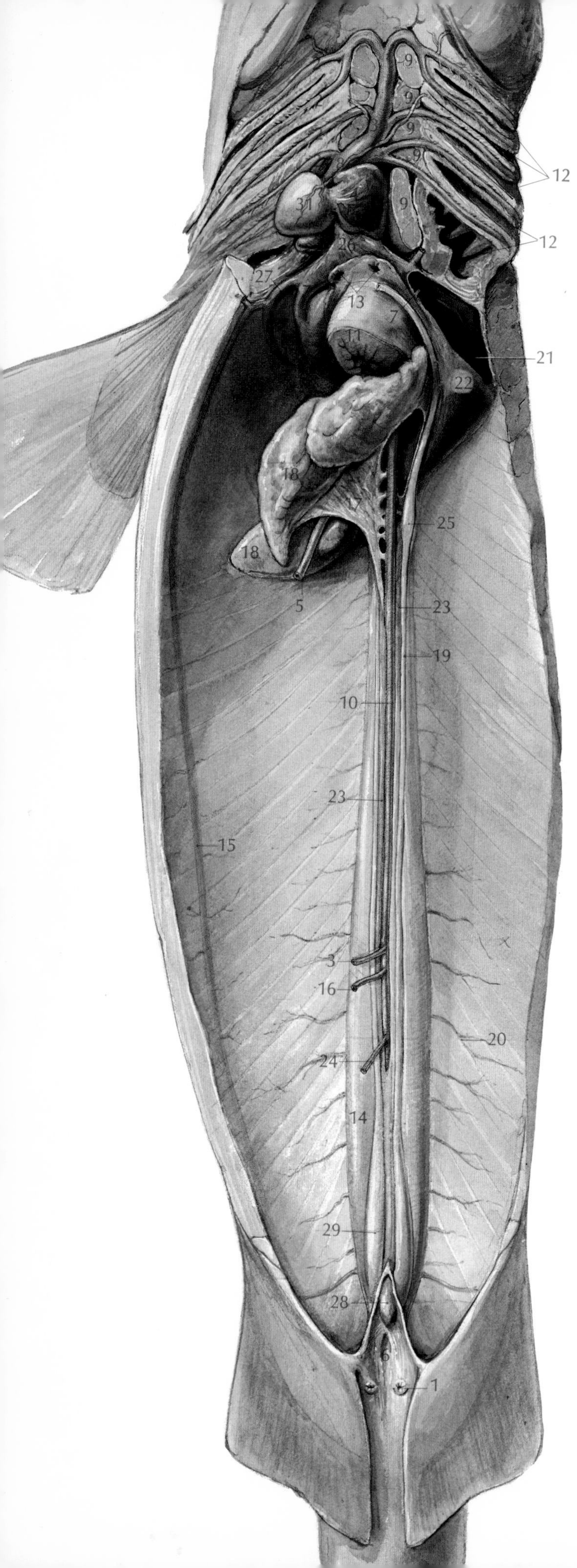


FIG. 15
THE POSTERIOR CARDINAL SINUSES AND VEINS

1 abdominal pore
2 afferent branchial artery
3 anterior mesenteric artery
4 atrium
5 celiac artery
6 cloacal opening
7 common cardinal vein
8 conus arteriosus
9 coracobranchial muscle
10 dorsal aorta
11 esophagus
12 gill pouches
13 hepatic sinuses opening into sinus venosus
14 kidney
15 lateral abdominal vein
16 lienogastric artery
17 mesovarium
18 ovary
19 oviduct
20 parietal vein
21 posterior cardinal sinus (cavity)
22 posterior cardinal sinus (lateral wall)
23 posterior cardinal vein
24 posterior mesenteric artery
25 shell gland
26 sinus venosus
27 subclavian vein
28 urinary papilla
29 uterus
30 ventral aorta
31 ventricle

FIG. 16
CROSS SECTIONS OF
THE TRUNK AND TAIL

1 caudal vein
2 dorsal aorta
3 kidney
4 oviduct
5 renal portal vein
6 right postcardinal vein
7 spinal cord
8 uterus

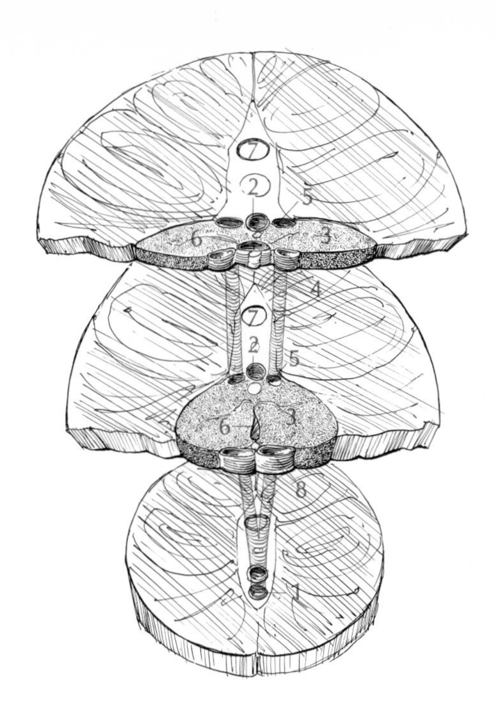

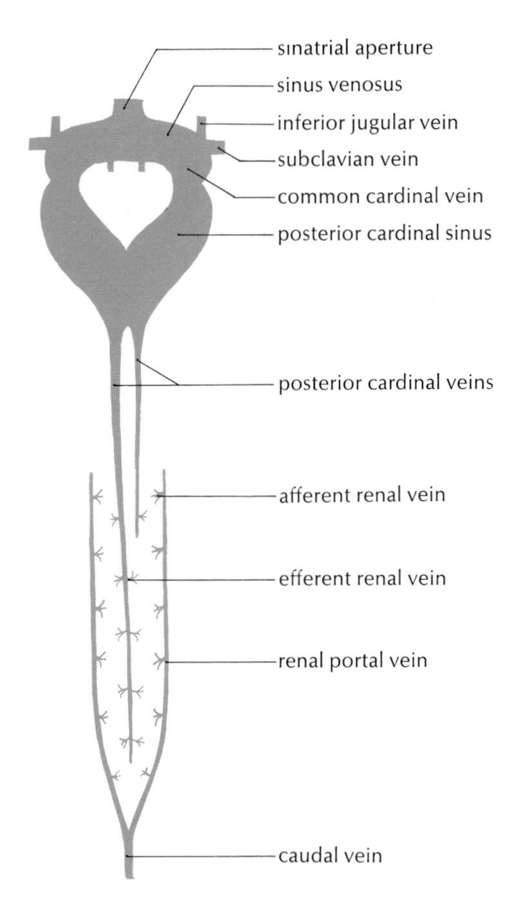


*veins of the trunk*

arches. Posteriorly the anterior cardinal sinus makes an abrupt ventral turn, joining the posterior cardinal sinus to form the common cardinal vein.

Remove the liver. Your dissection should now resemble Figure 15, except that you should not expose the ventral aorta and the afferent branchial arteries on the right as shown in this figure. Leave the gills and associated vessels intact on the right side for later study.

The posterior cardinal sinuses are large paired cavities which lie on the inner body wall dorsal to the esophagus. Anteriorly they are connected to the sinus venosus by the common cardinal veins, and posteriorly they are continuous with the narrow posterior cardinal veins. The right posterior cardinal vein is considerably longer and larger than the left; posteriorly it lies in the midline between the kidneys and receives efferent renal veins from both kidneys. The posterior cardinal sinuses receive tributaries from the body wall, the oviducts, and the gonads.

Refer to Figure 16. The renal portal system conveys venous blood from the tail to the capillaries of the kidneys; this blood is returned via the efferent renal veins to the posterior cardinal vein. The caudal vein is an unpaired vessel that lies within the haemal arch, ventral to the caudal artery (continuation of the dorsal aorta). At the posterior end of the kidneys the caudal vein divides into right and left renal portal veins, which extend anteriorly on the dorsal surface of each kidney. Blood from the renal portal veins enters the kidneys via numerous small afferent renal veins, which will not be visible unless the renal portal system is injected. Both the caudal and the renal portal veins receive segmental tributaries from the body wall. The caudal and renal portal veins can best be seen in a demonstration dissection of serial cross sections as illustrated in Figure 16.

The conus arteriosus extends as far as the anterior end of the

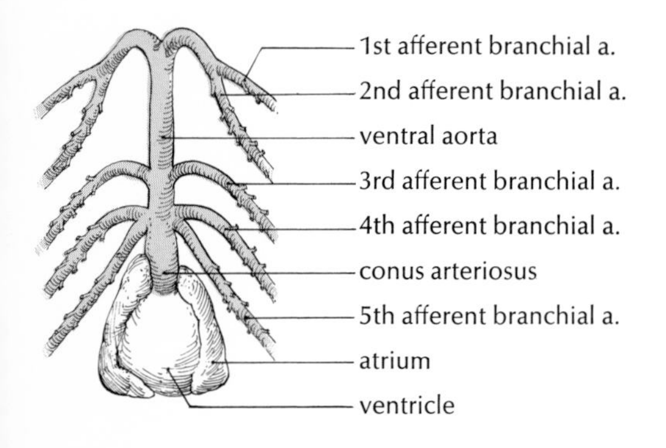


*the ventral aorta*

pericardial cavity; beyond this point it continues as the ventral aorta, a median vessel which gives off five pairs of afferent branchial arteries. See the ventral aorta and the afferent branchial arteries as illustrated in Figure 15. Follow the afferent branchial arteries on the left side of your specimen but leave the right side intact. Preserve the coronary arteries, which lie on the surface of the conus arteriosus and the ventricle.

The fourth and fifth afferent branchial arteries arise together just anterior to the pericardial cavity at the point where the conus arteriosus joins the ventral aorta. Shortly beyond this point the ventral aorta gives off the third afferent branchial artery. It then extends without further branches to the level of the hyoid arch, where it bifurcates, forming two stems which turn posteriorly. Each stem divides into two arteries, the first and second afferent branchials, which lie on either side of the first gill pouch. The first afferent branchial artery supplies the demibranch on the anterior aspect of the first gill pouch, and each of the other afferent branchial arteries supplies a holobranch. Observe the numerous small vessels given off from the afferent branchial arteries to the gill lamellae.

Cut through the angle of the jaw on the left side and continue the cut posteriorly through the gill arches. Pull the lower jaw open and secure it, exposing the oral cavity as illustrated in Figure 8, page 15. Identify the structures illustrated in Figure 8. Remove the mucous membrane from the roof of the mouth and identify the four pairs of efferent branchial arteries which converge in the midline to form the median dorsal aorta. Remove the gill cartilages on the intact right side, being careful to avoid injury to the branchial arteries and their branches. Then cut away the floor of the mouth to expose the dorsal aspect of the heart and the afferent branchial arteries. Your dissection should now resemble Figure 17, except that in this figure the branchial bars on the left side are intact and the liver has not been removed.

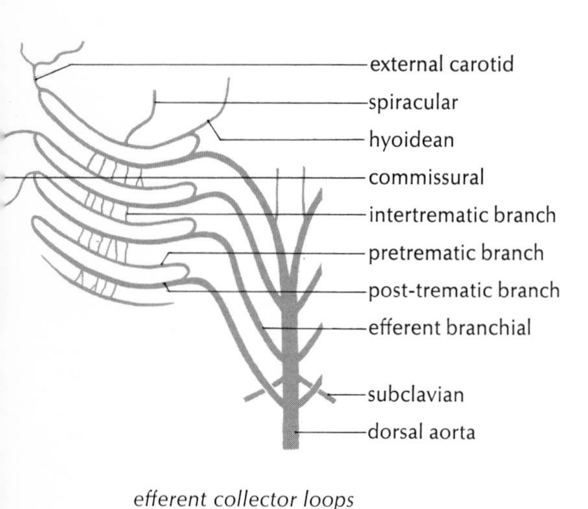


*efferent collector loops*

Unoxygenated blood from the heart is conveyed via the afferent branchial arteries to the gill lamellae, where the exchange of respiratory gases occurs. Each of the first four gill pouches is encircled by an efferent collector loop which returns oxygenated blood from the gill lamellae. Each loop consists of two branches: a large post-trematic branch posterior to the gill pouch, and a smaller pretrematic branch anterior to the gill pouch. Observe the numerous tributaries which convey blood from the lamellae to the pre- and post-trematic branches. Also observe that adjacent collector loops are connected to each other by several intertrematic branches. The fifth gill pouch has a demibranch on its anterior wall, but not on its posterior wall, and the corresponding collector loop is therefore incomplete.

The external carotid artery is a small vessel originating from the ventral end of the first collector loop. It passes ventral to the hyoid arch to supply the lateral aspect of the lower jaw.

The commissural artery originates from the ventral end of the second collector loop and passes posteriorly, ventral to the afferent branchial arteries, receiving branches from the third and sometimes the fourth collector loops. On the dorsal aspect of the conus arteriosus it anastomoses with the commissural artery of the opposite side, and then passes ventrally along the conus and onto the ventricle. A

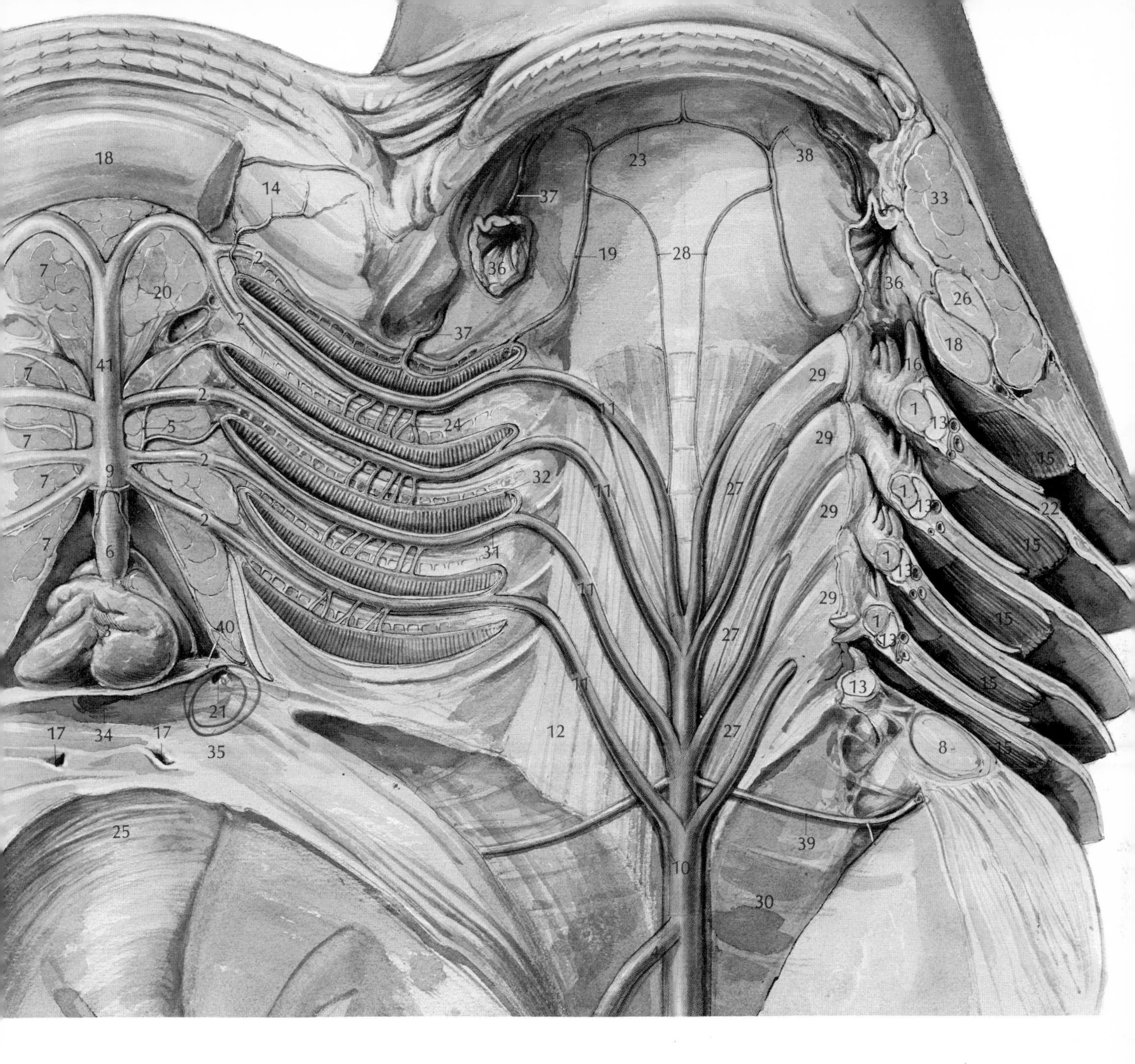


FIG. 17
THE BRANCHIAL ARTERIES

1 adductor muscle
2 afferent branchial artery
3 atrium
4 celiac artery
5 commissural artery
6 conus arteriosus
7 coracobranchial muscle
8 coracoid bar
9 coronary arteries
10 dorsal aorta
11 efferent branchial arteries
12 epaxial muscles
13 epibranchial cartilage
14 external carotid artery
15 gill lamellae
16 gill raker
17 hepatic sinus opening into sinus venosus
18 hyoid arch
19 hyoidean artery
20 hyoidean sinus
21 inferior jugular vein opening into sinus venosus
22 interbranchial septum
23 internal carotid artery
24 intertrematic branch
25 liver, dorsal surface
26 Meckel's cartilage
27 medial interarcual muscle
28 paired dorsal aortae
29 pharyngobranchial cartilage
30 posterior cardinal sinus
31 post-trematic branch
32 pretrematic branch
33 quadratomandibularis muscle
34 sinatrial aperture
35 sinus venosus
36 spiracle
37 spiracular artery
38 stapedial artery
39 subclavian artery
40 transverse septum
41 ventral aorta

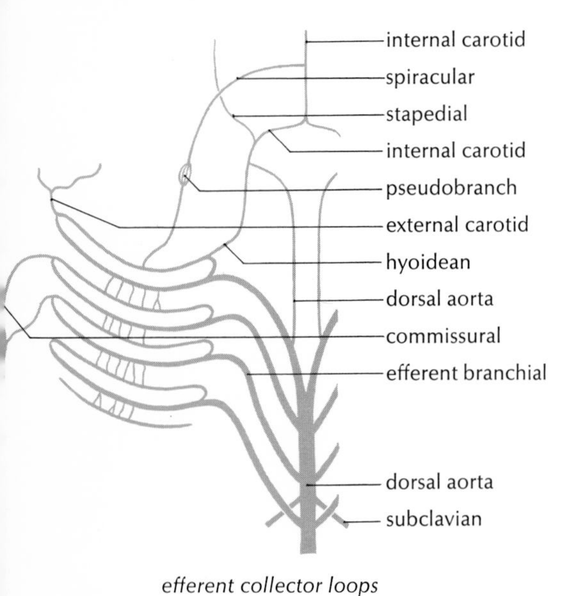

*efferent collector loops*

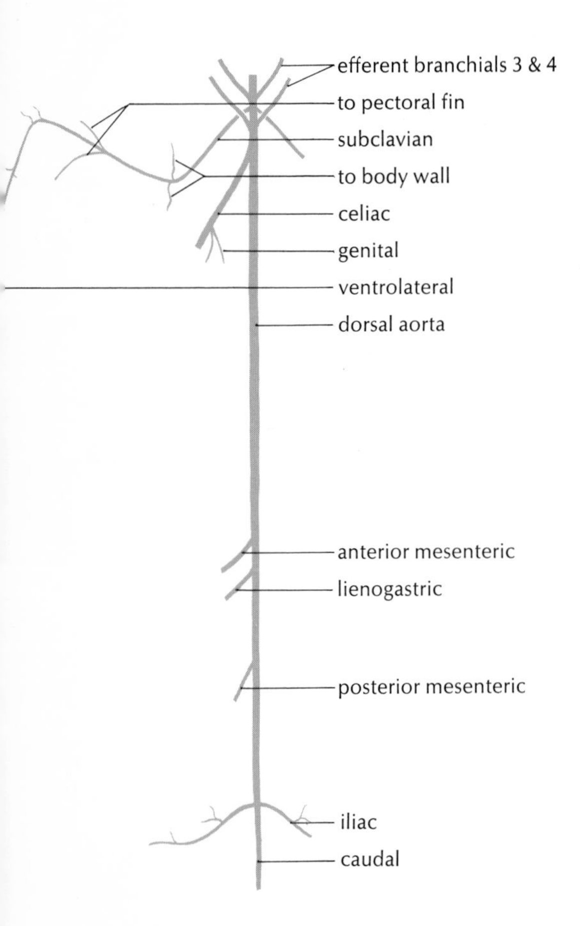

*branches of dorsal aorta*

branch of the commissural artery runs posteriorly along the lateral wall of the pericardial cavity to supply the pharynx and esophagus.

Near the middle of the pretrematic portion of the first collector loop the spiracular artery arises and passes anteriorly and dorsally, disappearing somewhat posterior to the spiracle. It supplies the pseudobranch (rudimentary lamellae within the spiracle) and then turns ventrally, reappearing anterior to the spiracle. Remove the mucous membrane and shave away the cartilage to find it at this point. It extends forward to join the internal carotid artery.

The hyoidean artery originates near the dorsal end of the first collector loop and passes anteriorly medial to the spiracle. It then divides into two branches: the stapedial artery, which turns laterally to supply the orbit, and the internal carotid, which turns medially and anastomoses with the internal carotid of the opposite side. The single artery formed by the union of the right and left internal carotids passes anteriorly to supply the brain, the eye, and the internal ear.

From the first efferent branchial arteries the slender paired dorsal aortae extend anteriorly on either side of the midline. Near the level of the spiracle they turn laterally to join the hyoidean arteries.

The subclavian artery originates from the dorsal aorta between the third and fourth efferent branchial arteries and passes laterally dorsal to the posterior cardinal sinus. It gives branches to the muscles of the dorsal body wall and then curves laterally and ventrally toward the pectoral fin, lying along the subscapular vein. It gives off two branches to the muscles at the base of the pectoral fin, one of which passes through a foramen in the coracoid bar. It then gives off a prominent posterior branch, the ventrolateral artery, which lies between the ventral abdominal vein and the midventral line of the body, supplying the muscles of the abdominal wall.

After giving off the subclavian artery, the dorsal aorta continues posteriorly along the dorsal body wall in the midline between the kidneys. In females it gives off paired oviducal arteries to the oviducts. The dorsal aorta gives off four unpaired arteries to the viscera: the celiac, the anterior mesenteric, the lienogastric, and the posterior mesenteric, which were identified during the dissection of the alimentary canal. Review pages 20-22. Also note that near its origin the celiac artery gives off genital branches to the gonads.

The dorsal aorta also gives off numerous paired parietal arteries to the body wall and renal arteries to the kidneys.

After giving off the posterior mesenteric artery, the dorsal aorta lies dorsal to the kidneys and is concealed from view. Near the cloaca it gives off paired iliac arteries to the cloaca and the pelvic fins, and continues into the tail as the caudal artery within the haemal canal.

Cut the ventral aorta just anterior to the point where the fourth and fifth afferent branchial arteries are given off, and remove the heart. Trim away the left lateral walls of the conus arteriosus, ventricle, and atrium as illustrated in Figure 18. The opening between the sinus venosus and the atrium is termed the sinatrial aperture; it is guarded by a pair of valves which prevent blood from returning from the atrium to the sinus venosus. Observe the very thin walls of the atrium, and the thin muscular strands on its inner surface. The opening between the atrium and the ventricle is termed the atrioventricular opening. It lies at the anterior end of the ventricle and to the right of the median line, and is guarded by a pair of pocket-shaped

FIG. 18
SAGITTAL SECTION
OF THE HEART

1 afferent branchial arteries 4 and 5
2 atrioventricular opening and valve
3 atrium
4 cavity of ventricle
5 common cardinal vein opening into sinus venosus
6 conus arteriosus
7 distal semilunar valve
8 hepatic sinus opening into sinus venosus
9 proximal semilunar valves
10 sinatrial aperture and valve
11 sinus venosus
12 wall of ventricle

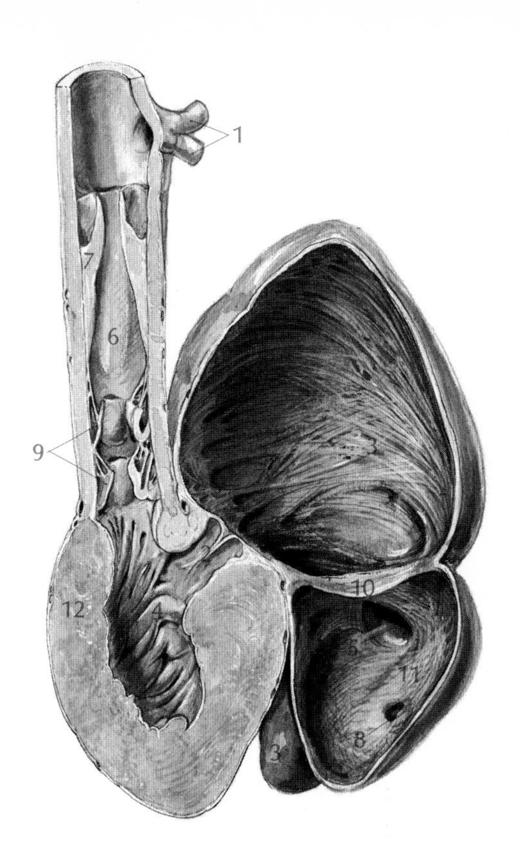


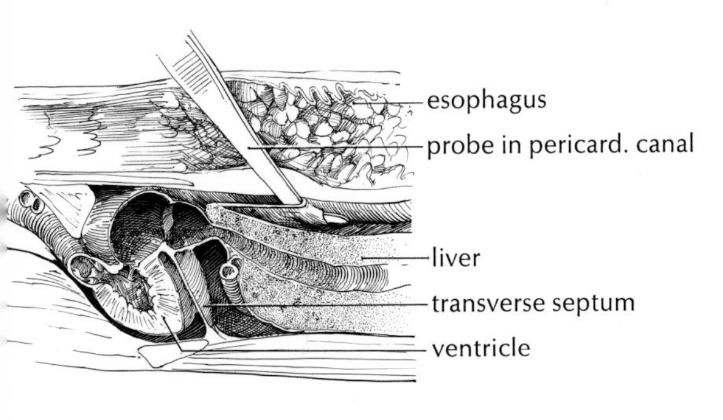


*sagittal section of heart, liver, and esophagus*

atrioventricular valves which prevent blood from returning from the ventricle to the atrium. Observe the thick, muscular walls of the ventricle, the relatively small ventricular cavity, and the irregular muscular ridges on the inner surface of the ventricle. On the inner surface of the conus arteriosus are nine semilunar valves, arranged in groups of three. The six proximal semilunar valves are quite small and lie at the posterior end of the conus. The three distal valves mark the anterior end of the conus and are considerably larger than the proximal valves. Probe the valves and examine them under water with a dissecting microscope.

Examine a sagittal section of the head and probe between the liver and the esophagus. The probe will enter a membranous pouch termed the pericardioperitoneal canal. Various authors give conflicting accounts of this canal. Goodrich describes it as Y-shaped, and says that its posterior end is open, forming a communication between the pericardial cavity and the pleuroperitoneal cavity, but O'Donoghue, who examined hundreds of specimens, found no such communication. The walls of the canal are delicate and easily torn, which makes it difficult to determine the position of the posterior opening, if there is one.

# THE UROGENITAL SYSTEM

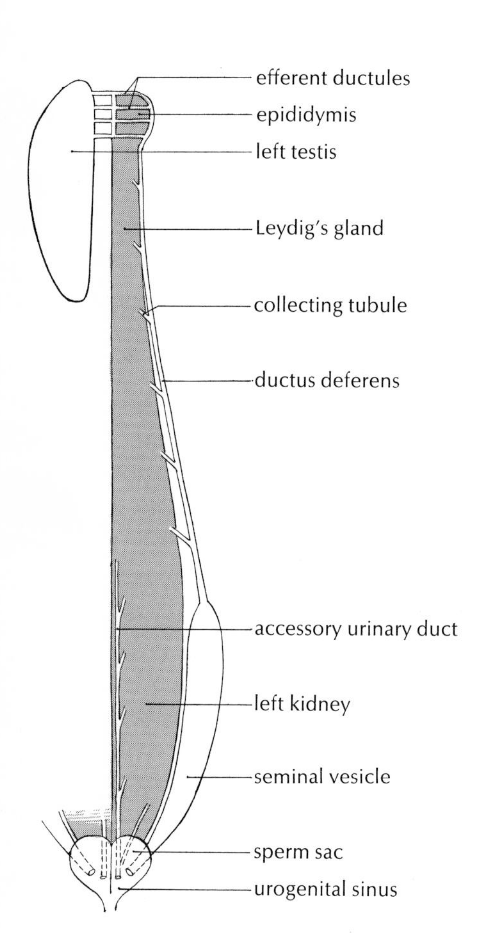


*male urogenital system*

Refer to Figure 19. The kidneys are slender organs that lie along the length of the dorsal body wall on either side of the midline. They are retroperitoneal, or behind the peritoneum. In the mature dogfish urine is produced only by the posterior portion of the kidney. In immature males urine is also produced by the anterior part of the kidney. In females the anterior part of the kidney is degenerate, having neither urinary nor reproductive functions.

In the shark the adrenal gland cannot be identified grossly as a separate structure, but its components are distributed along the medial borders of the kidneys as scattered cell clusters which can be identified by staining and microscopic examination.

If your specimen is a male, remove the peritoneum from the ductus deferens and the kidneys. In sexually immature specimens the ductus deferens is a straight, thin tube lying on the ventral surface of the kidney. In sexually mature specimens, the ductus deferens is intricately coiled, covering most of the ventral surface of the kidney. Spermatozoa pass from the testis through four small tubes termed efferent ductules located at the anterior end of the mesorchium, the mesenteric attachment of the testis. The efferent ductules are too small to be seen in gross dissection. From the efferent ductules the spermatozoa enter the anterior end of the kidney, termed the epididymis, in which the kidney tubules have been modified to serve as channels for the passage of the spermatozoa. They then enter the ductus deferens (Wolffian duct).

The ductus deferens receives collecting tubules from the anterior part of the kidney. In the immature male, urine passes from the kidney to the ductus deferens through these tubules. In the mature male, no urine is formed in the anterior part of the kidney (termed Leydig's gland); instead, the tubules transmit a milky contribution to the seminal fluid. Cut open the ductus deferens and observe the seminal fluid within it.

The posterior part of the ductus deferens is enlarged to form a wide tube termed the seminal vesicle, which receives no collecting tubules from the kidney. The seminal vesicle opens into a small receptacle termed the sperm sac, homolog of the posterior end of the oviduct. In the male, urine is collected from the kidney by a slender accessory urinary duct which lies along the medial side of

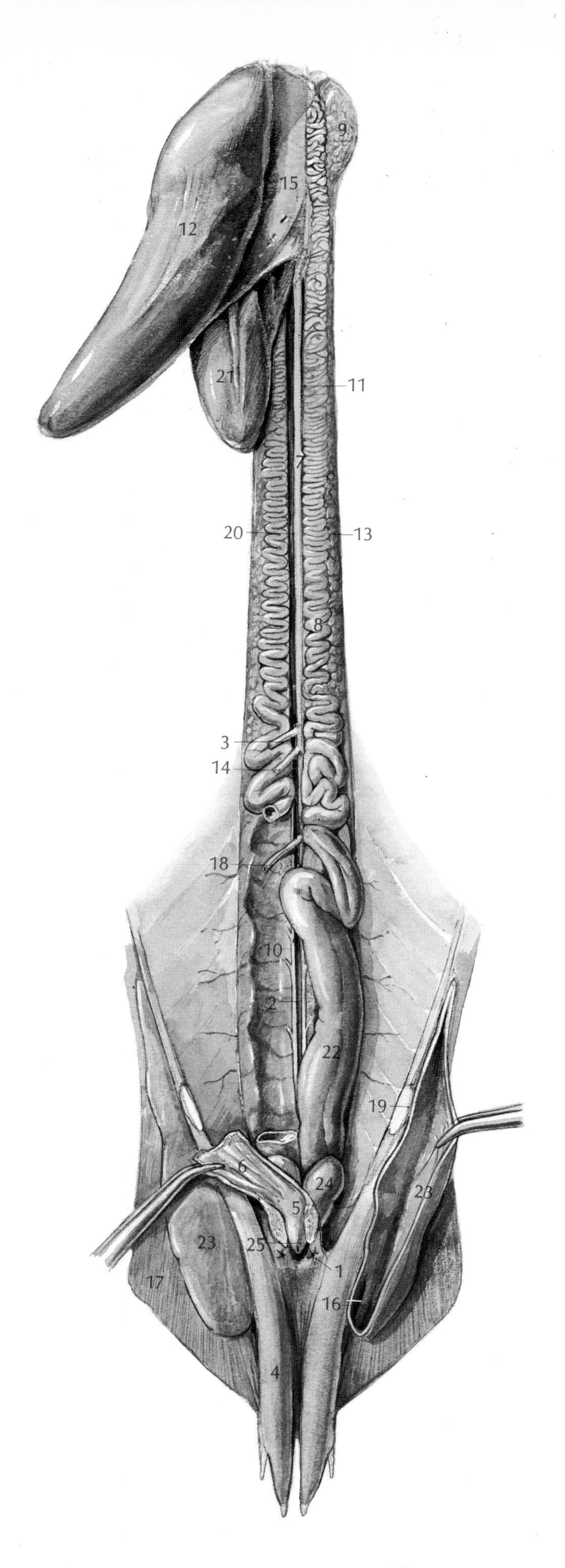


FIG. 19
THE MALE UROGENITAL SYSTEM

1 abdominal pore
2 accessory urinary duct
3 anterior mesenteric artery
4 clasper
5 cloaca
6 colon
7 dorsal aorta
8 ductus deferens
9 epididymis
10 kidney (opisthonephros)
11 left posterior cardinal vein
12 left testis
13 Leydig's gland
14 lienogastric artery
15 mesorchium
16 opening of siphon into clasper tube
17 pelvic fin
18 posterior mesenteric artery
19 puboischiac bar
20 right posterior cardinal vein
21 right testis
22 seminal vesicle
23 siphon
24 sperm sac
25 urogenital papilla

the kidney and opens into the sperm sac just medial and anterior to the opening of the seminal vesicle. The urogenital sinus is the part of the sperm sac posterior to the openings of the seminal vesicle and the accessory urinary duct. The right and left urogenital sinuses open by a common pore at the tip of the urogenital papilla, a small projection in the dorsal wall of the cloaca.

Remove the seminal vesicle from the right side as illustrated in Figure 19 and identify the accessory urinary duct and its collecting tubules (found in males only).

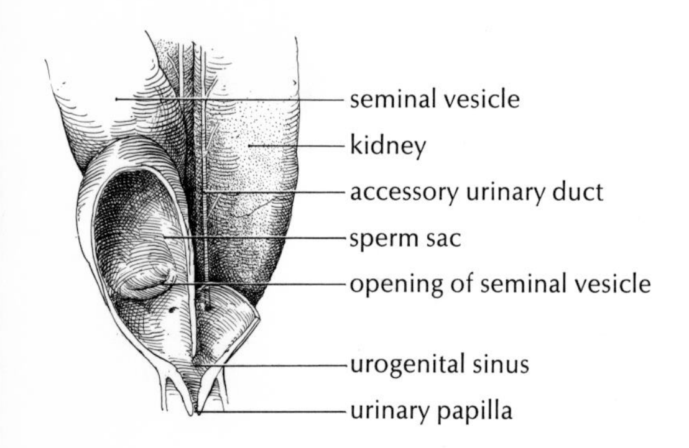


*sperm sacs, ventral view*

Remove the colon, slit open the urogenital papilla by making a cut along its ventral midline, and trim away the ventral walls of the sperm sacs to expose the inside of the sperm sacs as shown in the marginal diagram. The openings of the accessory urinary ducts are difficult to identify, but the seminal vesicles open into the sperm sacs by easily identified papillae.

The siphon is a thin-walled muscular sac which lies just under the skin on the ventral surface of the pelvic fin in the male. The clasper, found only in males, is a cylindrical modification of the inner lobe of the pelvic fin. Examine the medial surface of the clasper and observe that a groove, termed the clasper tube, runs along it. The cavity of the siphon is connected with the clasper tube by an aperture which lies at the posterior end of the siphon. Cut open the siphon and find this opening as illustrated in Figure 19.

The siphon secretes a substance which contributes to the seminal fluid and functions as a lubricant. During copulation the male inserts the clasper into the cloacal opening of the female and the siphon contracts, forcing fluid and spermatozoa through the clasper tube and into the cloaca of the female.

Observe the abdominal pores, two small openings on either side of the cloacal opening, just opposite the end of the urogenital papilla. Pass a probe through one of the abdominal pores and observe that it enters the pleuroperitoneal cavity. The function of the abdominal pores is obscure.

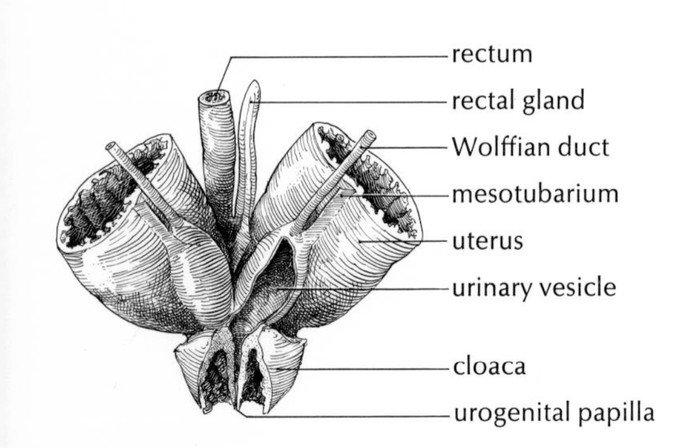


*urinary vesicle & uterus, dorsal view (after Marinelli & Strenger)*

In the female the anterior part of the kidney is degenerate and urine is produced only in the posterior part. Urine is collected by the Wolffian duct, a minute tube that lies just dorsal to the attachment of the mesotubarium and extends posteriorly along the ventral surface of the kidney to open by a pore at the tip of the urogenital papilla. Just anterior to this opening the Wolffian duct widens to form a small urinary vesicle. The Wolffian ducts are small and difficult to identify in females. They should not be confused with the tough white innominate ligament that lies between the kidneys in the midline.

In the female the Wolffian duct serves its primitive function of urine transport, and there is no accessory urinary duct, whereas in the male the Wolffian duct becomes the ductus deferens and serves as a passageway for seminal fluid and spermatozoa.

Refer to Figure 20. The ovaries lie on either side of the esophagus at the anterior end of the pleuroperitoneal cavity. Each ovary is suspended from the dorsal body wall by a mesentery termed the mesovarium.

The appearance of the ovaries varies with the sexual maturity of the specimen. In immature females the ovaries are relatively small and smooth. In mature females that have recently ovulated, the

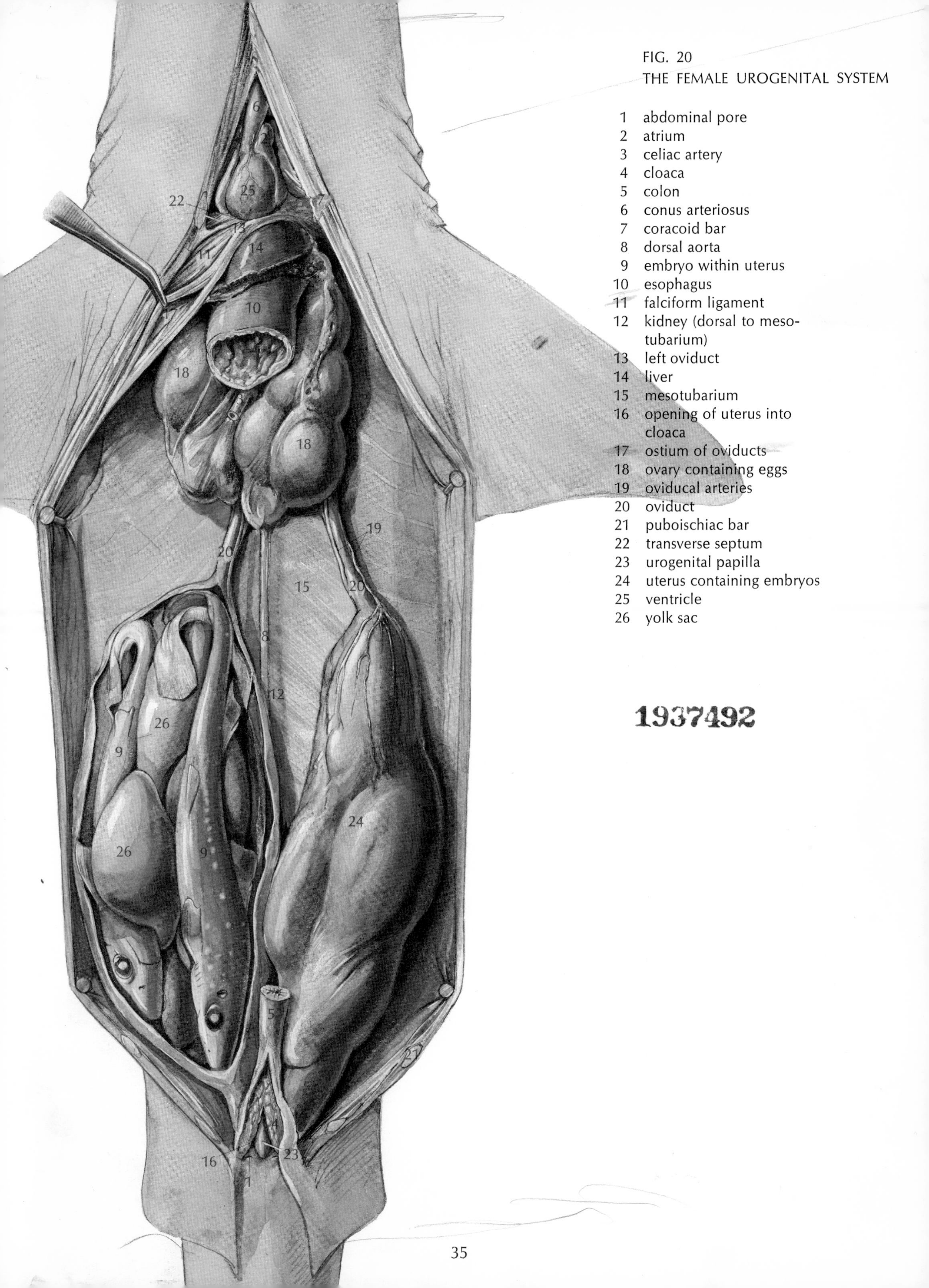


FIG. 20
THE FEMALE UROGENITAL SYSTEM

1 abdominal pore
2 atrium
3 celiac artery
4 cloaca
5 colon
6 conus arteriosus
7 coracoid bar
8 dorsal aorta
9 embryo within uterus
10 esophagus
11 falciform ligament
12 kidney (dorsal to mesotubarium)
13 left oviduct
14 liver
15 mesotubarium
16 opening of uterus into cloaca
17 ostium of oviducts
18 ovary containing eggs
19 oviducal arteries
20 oviduct
21 puboischiac bar
22 transverse septum
23 urogenital papilla
24 uterus containing embryos
25 ventricle
26 yolk sac

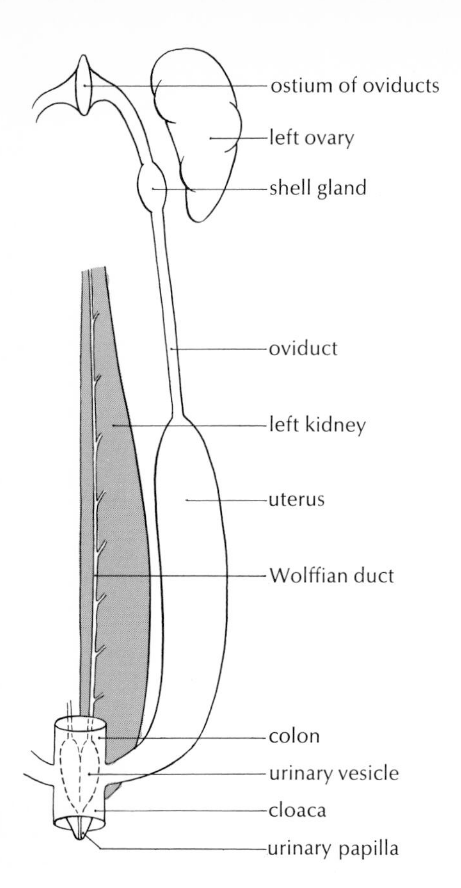


*female urogenital system*

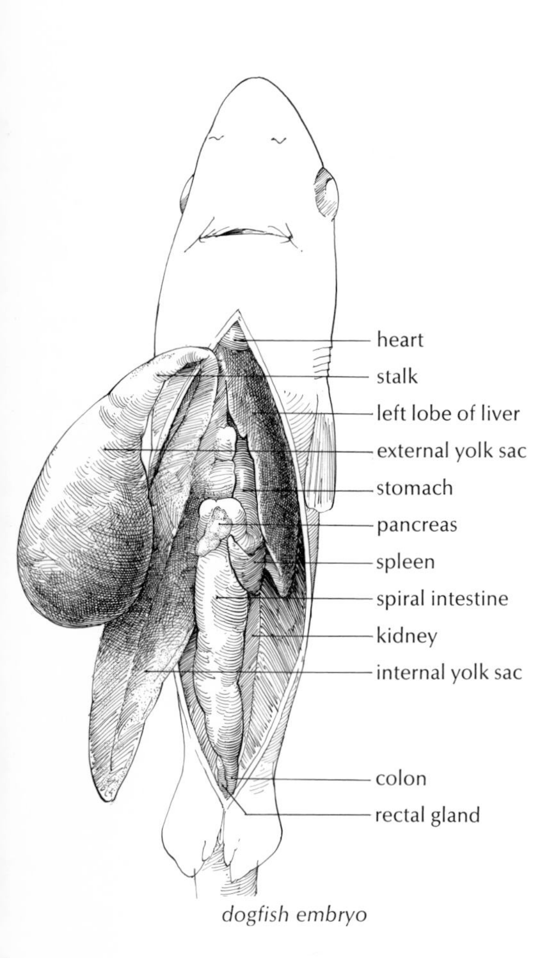


*dogfish embryo*

ovaries contain numerous small developing ova. If fertilization has occurred (as is almost always the case), the uterus contains two to four small embryos. As the period of gestation progresses, two to four of the ova in each ovary grow to large size and assume a yellowish hue because of the large amount of yolk they contain, while at the same time the other ova are resorbed by the ovary.

The oviducts open into the pleuroperitoneal cavity by a common ostium which lies within the falciform ligament, the ligament that attaches the liver to the ventral body wall. Just anterior to the ostium the right and left oviducts separate and pass around the anterior end of the liver, extending posteriorly along the ventral surface of the kidneys, to which they are attached by a mesentery termed the mesotubarium. Posteriorly they open into the cloaca. Near the anterior end of the oviduct is an enlargement termed the shell gland, which serves as a receptacle for spermatozoa; the shell gland also secretes a membrane that encloses the eggs as they pass through it. Posteriorly the oviduct enlarges to form the uterus, within which the embryonic shark develops.

The appearance of the oviducts varies greatly, depending on the maturity of the specimen. In immature females the ostium is small and difficult to find, and the oviducts are slender, straight tubes lying close to the ventral surface of the kidney. In mature females the ostium is a wide opening and the oviduct and uterus are much larger and are suspended from the ventral surface of the kidney by a well-developed mesotubarium. The immature female is illustrated in Figure 15, page 26; the pregnant female is illustrated in Figure 20.

In ovulation the ova break through the wall of the ovary and pass into the pleuroperitoneal cavity, from which they are moved to the ostium of the oviduct by the action of peritoneal cilia. They pass posteriorly through the oviduct to the shell gland, within which spermatozoa may be stored for some time. In the shell gland the ova are fertilized and enclosed in a protective membrane. The membrane and the eggs together constitute a structure termed the candle. Each candle usually contains two to four embryos. The candle passes into the uterus, and as the embryos develop the protective membrane is absorbed.

The period of gestation is surprisingly long—twenty to twenty-two months. The developing dogfish lies freely within the uterus. Virtually all the nourishment consumed by the young dogfish during its two years of prenatal life is contained in the yolk sac.

Cut open the pleuroperitoneal cavity of an embryonic dogfish and observe that the stalk of the external yolk sac is continuous with an internal yolk sac which is connected directly to the alimentary canal. The ciliated linings of the external and internal yolk sacs move the yolk into the alimentary canal, where it is digested. At birth the nutriment in the external yolk sac has been consumed, but some of the internal yolk sac still remains.

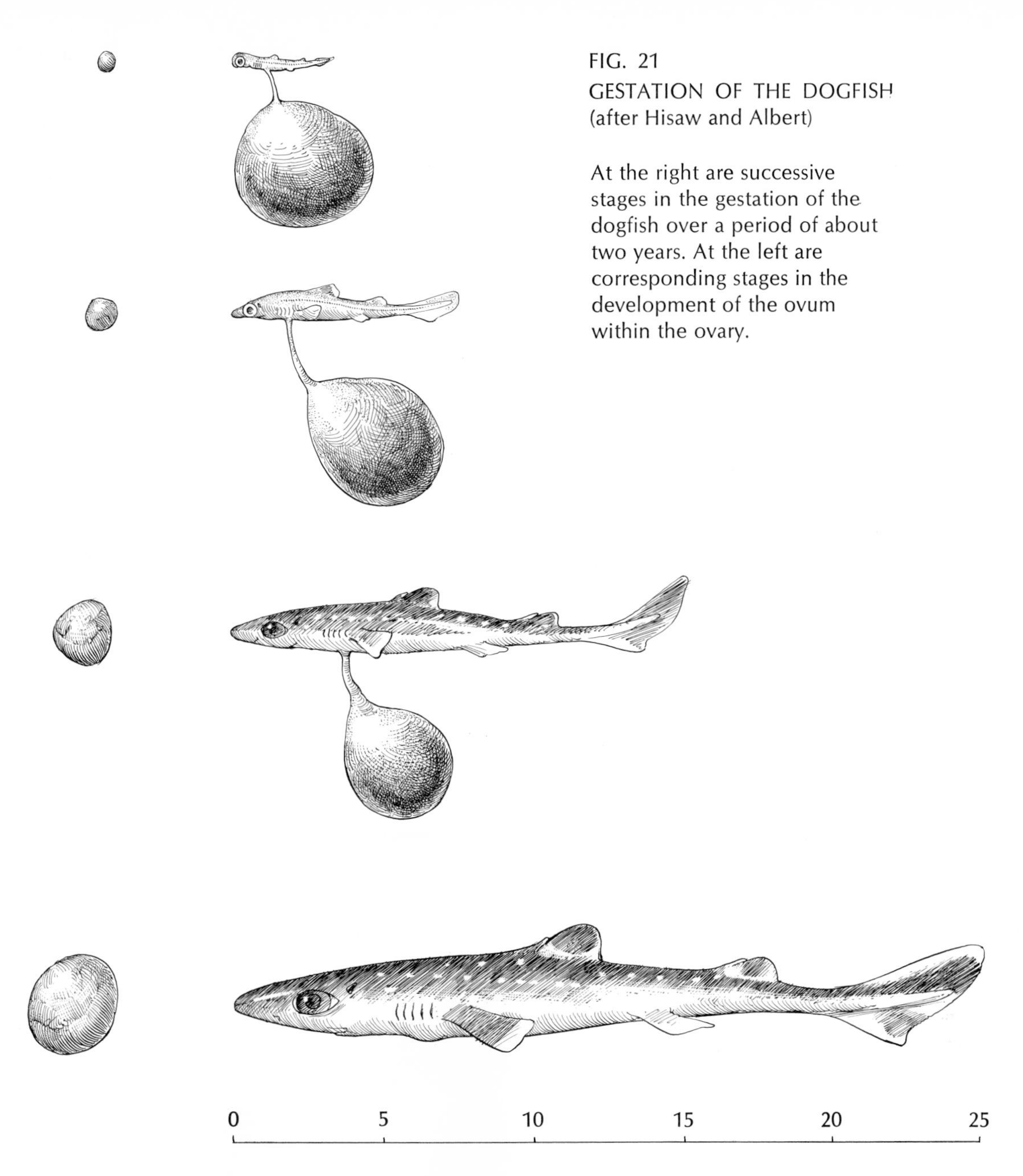


FIG. 21
GESTATION OF THE DOGFISH
(after Hisaw and Albert)

At the right are successive stages in the gestation of the dogfish over a period of about two years. At the left are corresponding stages in the development of the ovum within the ovary.

# THE NERVOUS SYSTEM

The instructor should select certain students to remove the dorsal part of the chondrocranium, exposing the structures illustrated in Figure 23; other students should remove the lower jaw and the ventral part of the chondrocranium to expose the structures illustrated in Figure 24. For this exercise new extra large *Squalus* heads should be provided.

In Figures 23 and 24 the membranous labyrinth is illustrated schematically within the otic capsule. In the dissection of the dorsal aspect, remove the otic capsule on the right to expose the glossopharyngeal nerve, which lies ventral to it. On the left leave the otic capsule intact for later study. In the dissection of the ventral aspect, leave the otic capsules intact on both sides.

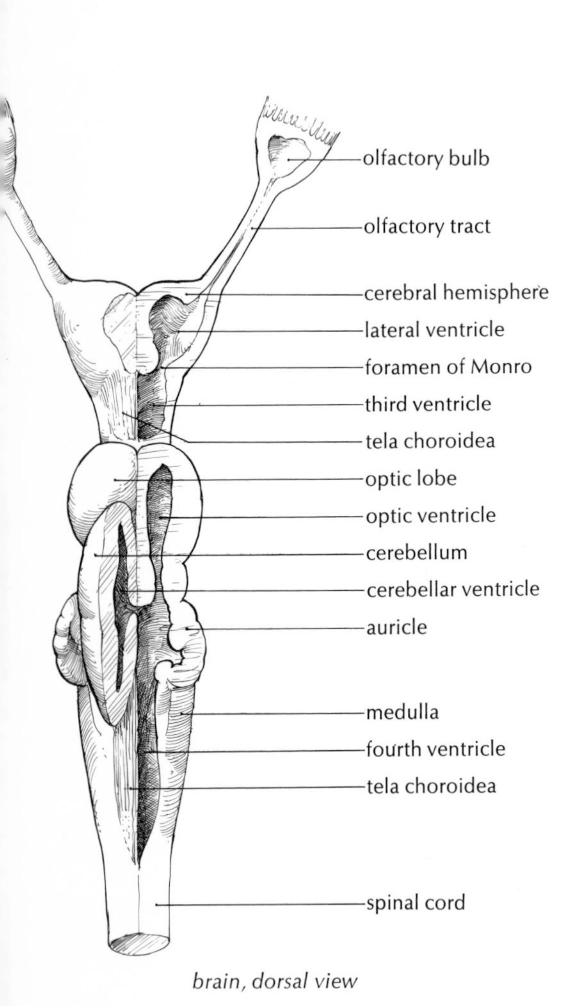


*brain, dorsal view*

Familiarize yourself with the divisions of the brain as summarized in Figure 22. Although these divisions do not in every case represent structural or functional entities, they do provide a convenient system for identifying homologous areas in the brains of various vertebrates, and your study of the brain will be much simplified if you master these terms at the outset.

The brain and spinal cord are enclosed in a delicate vascular membrane termed the primitive meninx. Strands of tissue connect the primitive meninx to the membrane lining the inner wall of the cranial cavity. Examine the primitive meninx by pulling it away from the dorsal surface of the brain with small forceps. In life cerebrospinal fluid fills the space between the brain and the wall of the cranial cavity.

Begin your study of the brain by examining a dissection of the dorsal aspect as illustrated in Figure 23.

On the right side identify the olfactory sac, a roughly spherical structure into which the external naris opens. Cut into the sac and observe that it contains radial folds, termed lamellae, which are covered by olfactory epithelium. There is no communication between the cavity of the olfactory sac and the oral cavity. The structure of the lamellae can best be seen in a cross section like that illustrated in the marginal diagram on page 53.

The telencephalon is the most anterior part of the brain. It consists of the olfactory bulbs, the olfactory tracts, and the cerebral hemispheres.

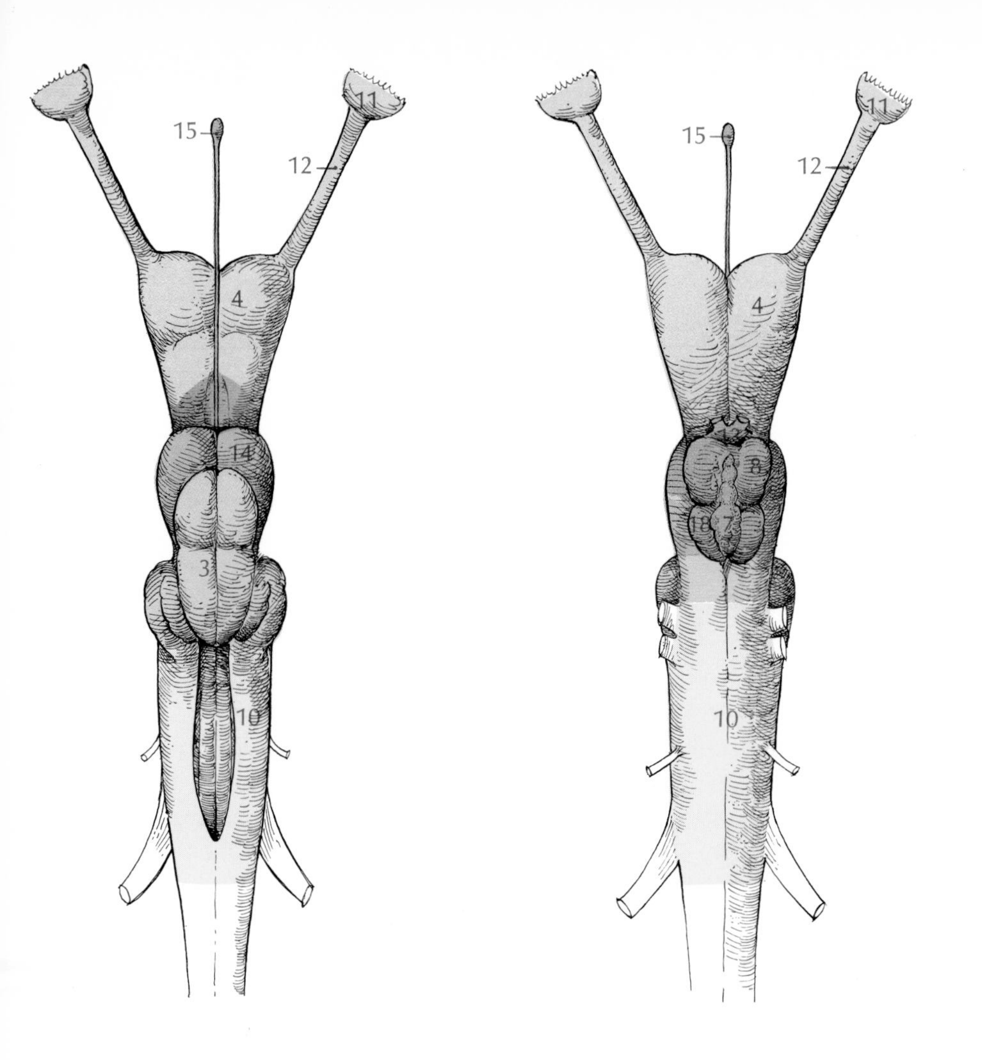


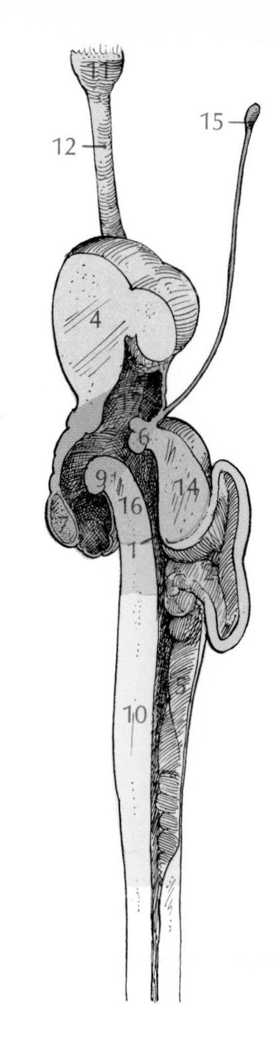


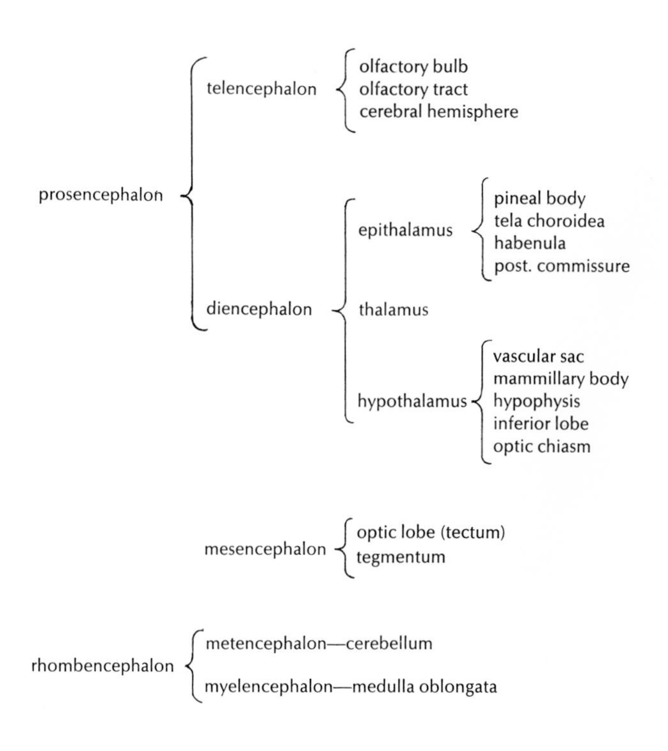


FIG. 22
DIVISIONS OF THE BRAIN

1 aqueduct
2 cerebellar ventricle
3 cerebellum
4 cerebral hemisphere
5 fourth ventricle
6 habenula and posterior commissure
7 hypophysis
8 inferior lobe of infundibulum
9 mammillary body
10 medulla oblongata
11 olfactory bulb
12 olfactory tract
13 optic chiasm
14 optic lobe
15 pineal body
16 tegmentum
17 third ventricle and thalamus
18 vascular sac

blue telencephalon
purple diencephalon
green mesencephalon
red metencephalon
yellow myelencephalon

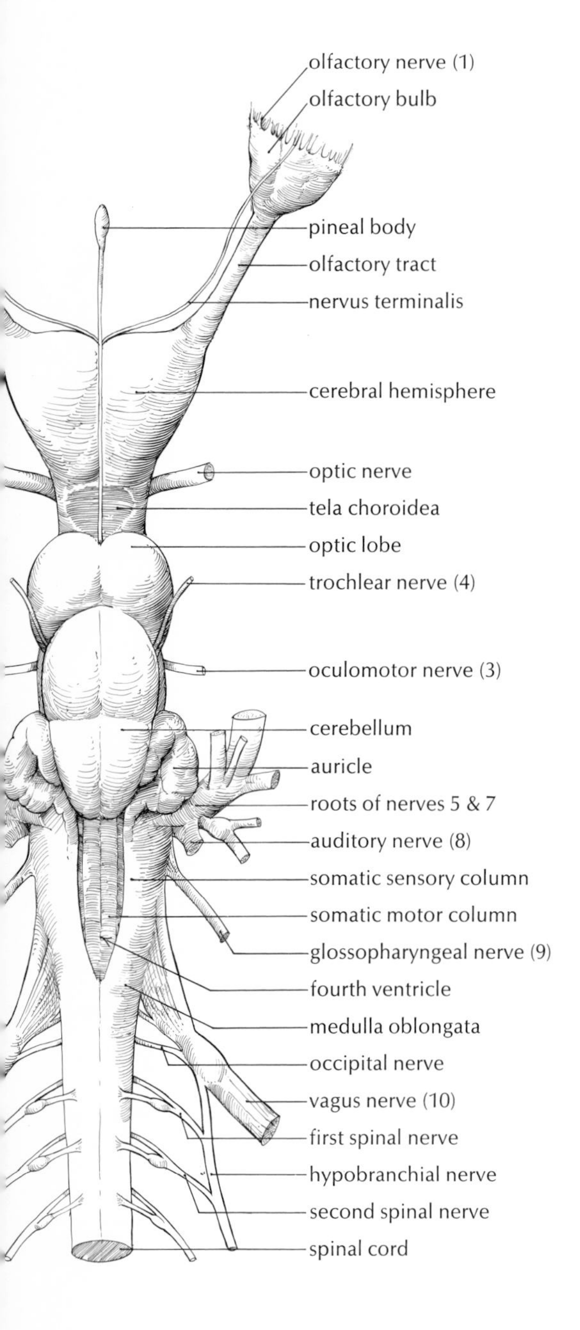


The olfactory bulb lies on the medial wall of the olfactory sac. The olfactory tract is an outgrowth of the brain which connects the olfactory bulb with the cerebral hemisphere. The cerebral hemispheres are paired oval lobes which form the anterior part of the main body of the brain. Each hemisphere contains a cavity termed a lateral ventricle, which is continuous with the central cavity of the olfactory bulb and tract. Posteriorly each lateral ventricle communicates, via an opening termed the foramen of Monro, with the unpaired median third ventricle.

Posterior to the cerebral hemispheres on the dorsal aspect of the brain is a thin membrane covering a cavity. Remove the membrane and probe the cavity. This is the cavity of the third ventricle. The diencephalon is the part of the brain immediately surrounding the third ventricle. The membranous roof of the third ventricle is the tela choroidea. It is formed by a fusion of the primitive meninx and the ependyma, a membrane which lines the central canal of the spinal cord and the cavities of the brain. Vascular tufts termed the anterior choroid plexuses extend from the tela choroidea into the third ventricle and the lateral ventricles.

The diencephalon consists of three parts: the epithalamus, dorsal to the third ventricle; the thalamus, an area of gray matter in the lateral wall of the third ventricle; and the hypothalamus, ventral to the third ventricle. The components of the epithalamus will be seen in the sagittal section of the brain; the components of the hypothalamus will be seen in the ventral view and in the sagittal section.

From the posterior margin of the diencephalon a slender strand of tissue extends anteriorly, ending in the epiphysial foramen in the roof of the chondrocranium. This is the epiphysis, or pineal body, which is part of the epithalamus. It is torn when the roof of the chondrocranium is removed, and should be observed in the sagittal section of the complete head. It is the degenerate rudiment of a median third eye found in many primitive extinct vertebrates.

The paired bulges posterior and dorsal to the diencephalon are the optic lobes, or optic tectum. They form the dorsal part of the mesencephalon. Each optic lobe contains an optic ventricle which communicates with the cerebral aqueduct, the central cavity of the mesencephalon.

The oval structure dorsal and posterior to the optic lobes is the cerebellum, the dorsal part of the metencephalon. It contains a cavity termed the cerebellar ventricle, which communicates ventrally with the aqueduct and the fourth ventricle. Faint grooves, lying at right angles to each other, divide the cerebellum into four segments. At the posterior end of the cerebellum are two lateral projections termed the auricles of the cerebellum.

The myelencephalon, or medulla oblongata, is the oblong part of the brain between the cerebellum and the spinal cord. The cavity of the medulla is the fourth ventricle, which communicates anteriorly with the aqueduct and posteriorly with the central canal of the spinal cord. Like the third ventricle, the cavity of the fourth ventricle is roofed over dorsally by a tela choroidea, and from this a choroid plexus extends into the fourth ventricle. Remove the tela choroidea and probe the cavity of the fourth ventricle.

Two parallel ridges can be seen in the floor of the fourth ventricle. These are the somatic motor columns, the most ventrally placed of

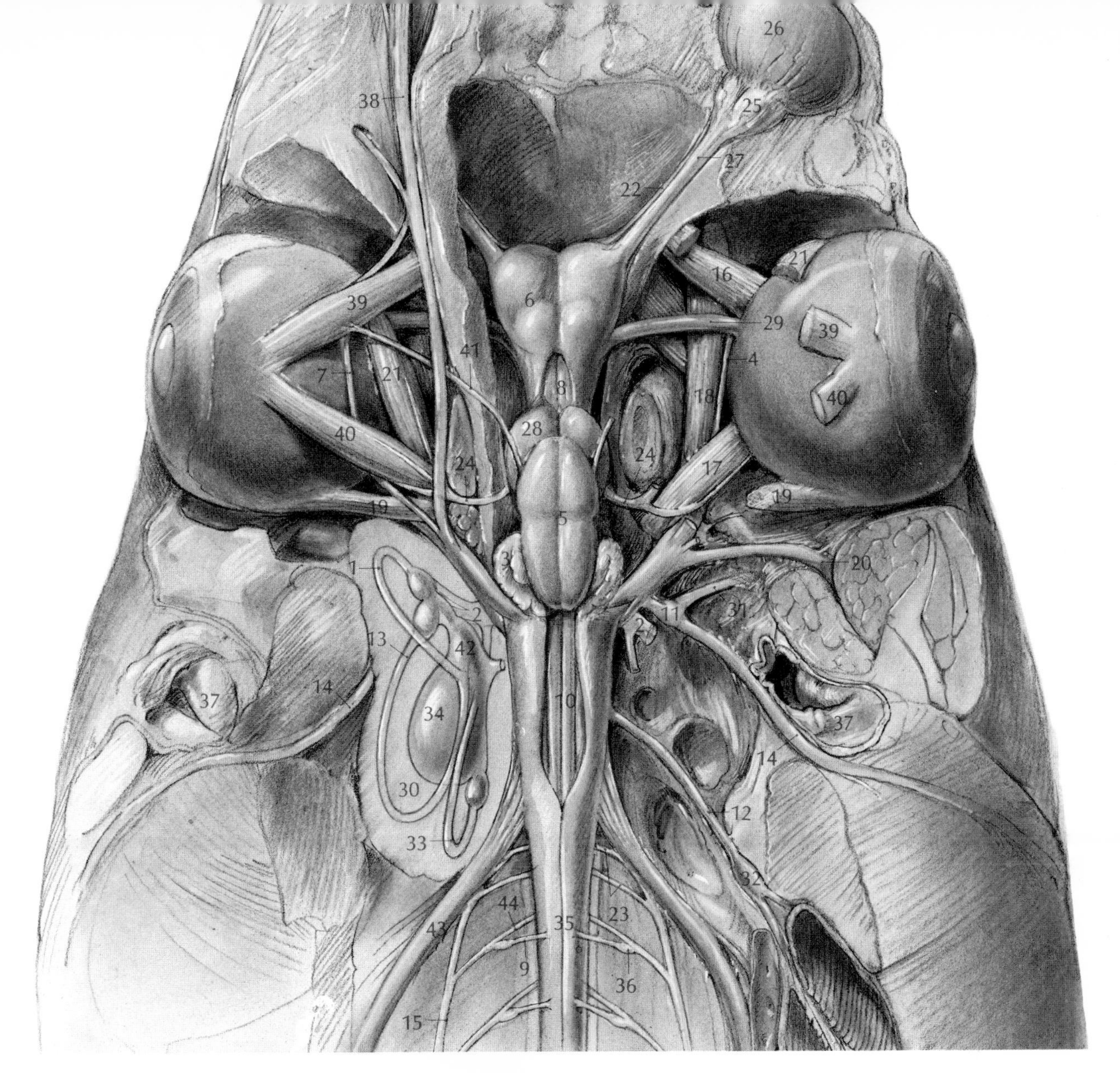

FIG 23
DORSAL VIEW OF THE BRAIN AND CRANIAL NERVES

On the right the following structures are removed: otic capsule, membranous labyrinth, superior oblique, superior rectus, lateral rectus, superficial ophthalmic trunk and optic pedicle.

1 anterior semicircular duct
2 auditory nerve (8)
3 auricle of cerebellum
4 branch of oculomotor nerve
5 cerebellum
6 cerebral hemisphere
7 deep ophthalmic nerve (5)
8 diencephalon and third ventricle
9 dorsal root of first spinal nerve
10 fourth ventricle and medulla oblongata
11 geniculate ganglion
12 glossopharyngeal nerve (9)
13 horizontal semicircular duct
14 hyomandibular trunk (7)
15 hypobranchial nerve
16 inferior oblique muscle
17 inferior rectus muscle
18 infraorbital trunk (5, 7)
19 lateral rectus muscle
20 mandibular nerve (5)
21 medial rectus muscle
22 nervus terminalis
23 occipital nerves
24 oculomotor nerve (3)
25 olfactory bulb
26 olfactory sac
27 olfactory tract
28 optic lobe
29 optic nerve (2)
30 otic capsule
31 palatine nerve (7)
32 petrosal ganglion
33 posterior semicircular duct
34 sacculus
35 spinal cord
36 spinal ganglion
37 spiracle
38 superficial ophthalmic trunk (5, 7)
39 superior oblique muscle
40 superior rectus muscle
41 trochlear nerve (4)
42 utriculus
43 vagus nerve (10)
44 ventral root of spinal nerve

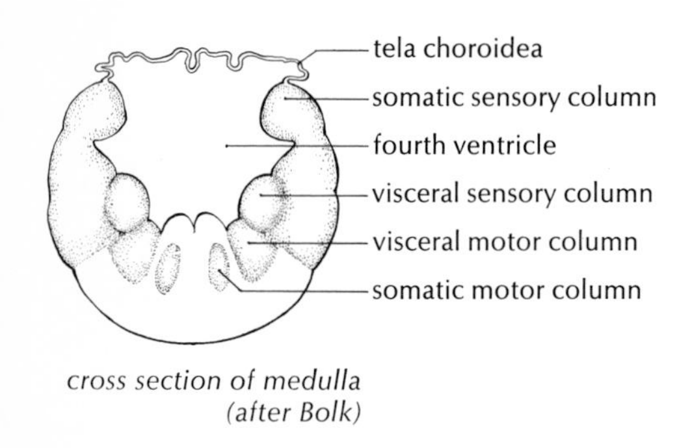


*cross section of medulla (after Bolk)*

*brain, ventral view*

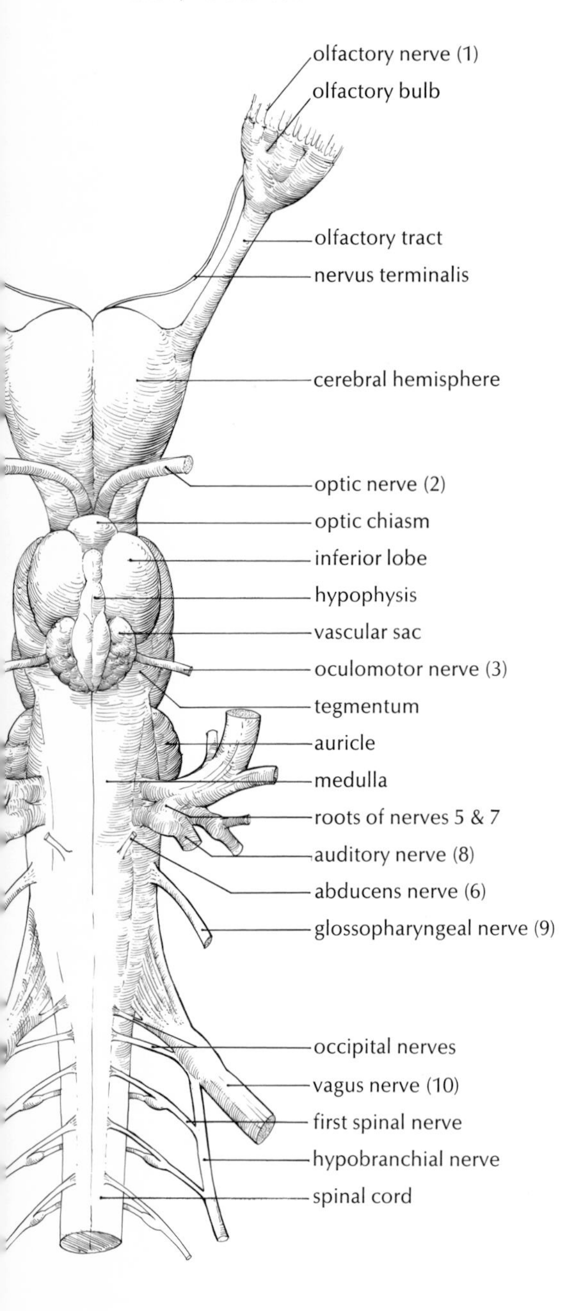


four functional columns of gray matter in the medulla. Lateral to the somatic motor column is the visceral motor column. These columns contain the cell bodies of somatic motor and visceral motor neurons which constitute the motor components of cranial nerves originating in the medulla. Dorsal to the motor columns are the visceral sensory and somatic sensory columns, which receive and relay sensory impulses from cranial nerves originating in the medulla. See the two sensory columns as illustrated in the sagittal section of the brain, Figure 27. The visceral sensory column is divided into several rounded elevations, each of which is supposed to correspond to one of the gill arches. At the anterior end of the somatic sensory column is an elevation termed the acousticolateral area, which receives and relays impulses from the membranous labyrinth and the lateral line canals. The sensory and motor columns are continuous posteriorly with corresponding columns in the spinal cord. Anteriorly they continue into the mesencephalon, where they break up, forming individual nuclei of gray matter.

Now examine a specimen in which the ventral surface of the brain has been exposed. Identify the olfactory bulbs, olfactory tracts, and cerebral hemispheres (parts of the telencephalon).

The hypothalamus (the ventral part of the diencephalon) is seen posterior to the cerebral hemispheres. The parts of the hypothalamus visible in the ventral view are the optic chiasm, the hypophysis (pituitary body), and the infundibulum. The optic chiasm is the point at which the optic nerves cross. In all vertebrates except mammals, all the fibers from the optic nerve of the right eye go to the optic lobe on the left side of the brain, and vice versa. In the lateral view of the brain the course of the fibers passing from the optic chiasm to the optic lobe can be seen as a band termed the optic tract (Fig. 27). The infundibulum consists of paired inferior lobes, which lie just posterior to the optic chiasm, and of the vascular sac, a thin-walled sac posterior to the inferior lobes.

In the midline ventral to the vascular sac and the inferior lobes of the infundibulum is an elongated structure termed the hypophysis or pituitary body. It is an endocrine gland formed by the fusion of the ventral part of the infundibulum and a dorsal evagination, the hypophyseal pouch, which arises from the roof of the mouth in the embryo.

The floor of the mesencephalon lies dorsal to the hypophysis. It is termed the tegmentum. Posterior to it are the floor of the metencephalon and the floor of the myelencephalon. There is no externally visible dividing line between these two parts.

Begin your study of the cranial nerves by familiarizing yourself with their names and functions as summarized in Figure 25. The functions of the cranial nerves are similar in all vertebrates, and a good understanding of the simplified scheme presented in this figure will provide a basic orientation for a more detailed examination of the cranial nerves in the shark and in higher vertebrates. For the sake of simplicity I have not described the visceral and somatic components of the cranial nerves; they will be covered in your general text. As you read the descriptions of each cranial nerve on the following pages, refer to Figures 23 and 24, and to several dissections from both dorsal and ventral aspects, identifying the nerves as far as possible in both views.

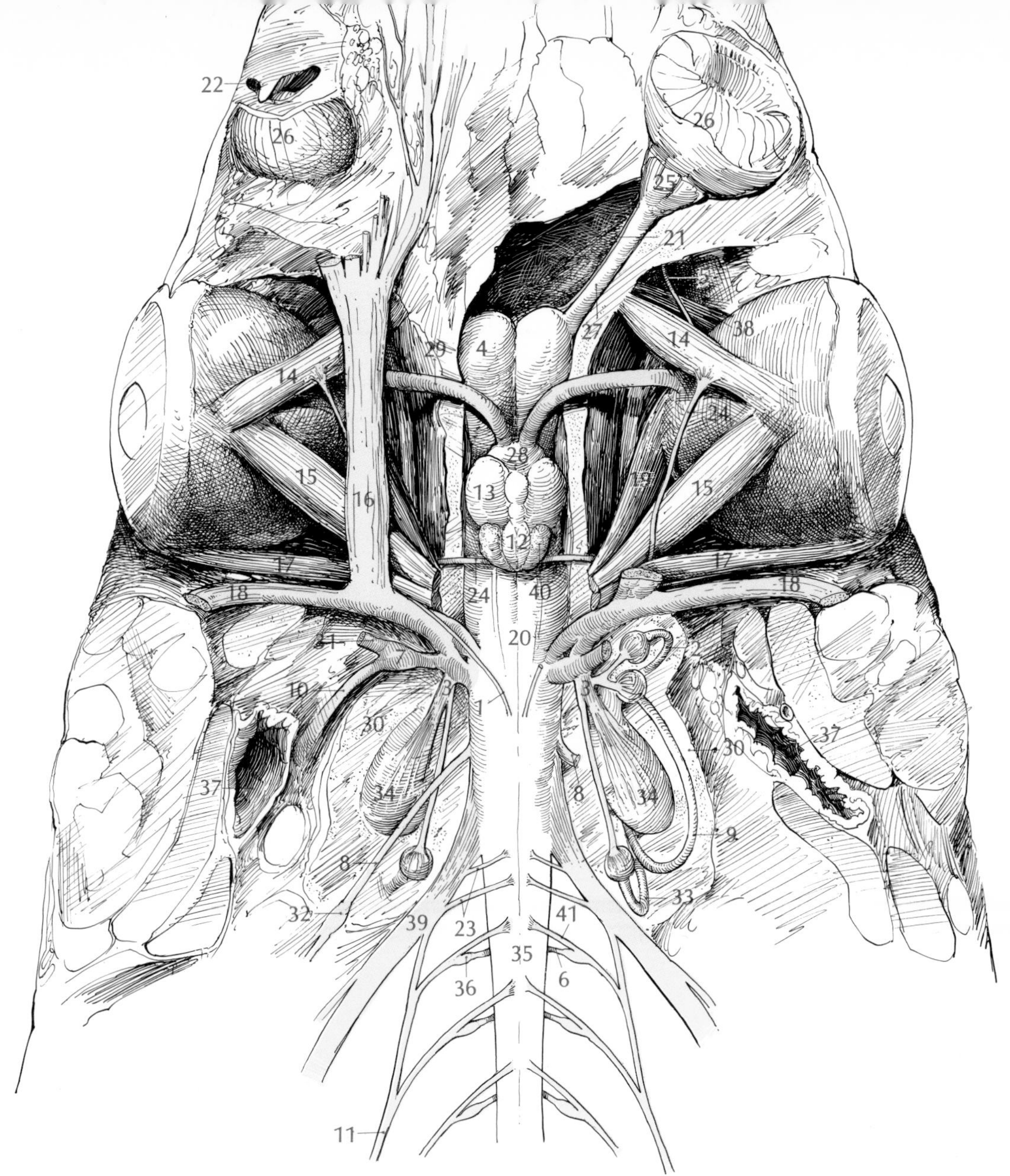


FIG. 24
VENTRAL VIEW OF THE BRAIN AND CRANIAL NERVES

On the left the following structures are removed:
infraorbital trunk, geniculate ganglion, palatine nerve,
hyomandibular trunk, glossopharyngeal nerve, and abducens nerve

1 abducens nerve (6)
2 anterior semicircular duct
3 auditory nerve (8)
4 cerebral hemisphere
5 deep ophthalmic nerve (5)
6 dorsal root of first spinal nerve
7 geniculate ganglion
8 glossopharyngeal nerve (9)
9 horizontal semicircular duct
10 hyomandibular trunk (7)
11 hypobranchial nerve
12 hypophysis
13 inferior lobe of infundibulum
14 inferior oblique muscle
15 inferior rectus muscle
16 infraorbital trunk (5, 7)
17 lateral rectus muscle
18 mandibular nerve (5)
19 medial rectus muscle
20 medulla oblongata
21 nervus terminalis
22 nostril
23 occipital nerves
24 oculomotor nerve (3)
25 olfactory bulb
26 olfactory sac
27 olfactory tract
28 optic chiasm
29 optic nerve (2)
30 otic capsule
31 palatine nerve (7)
32 petrosal ganglion
33 posterior semicircular duct
34 sacculus
35 spinal cord
36 spinal ganglion
37 spiracle
38 superior oblique muscle
39 vagus nerve (10)
40 vascular sac
41 ventral root of first spinal nerve

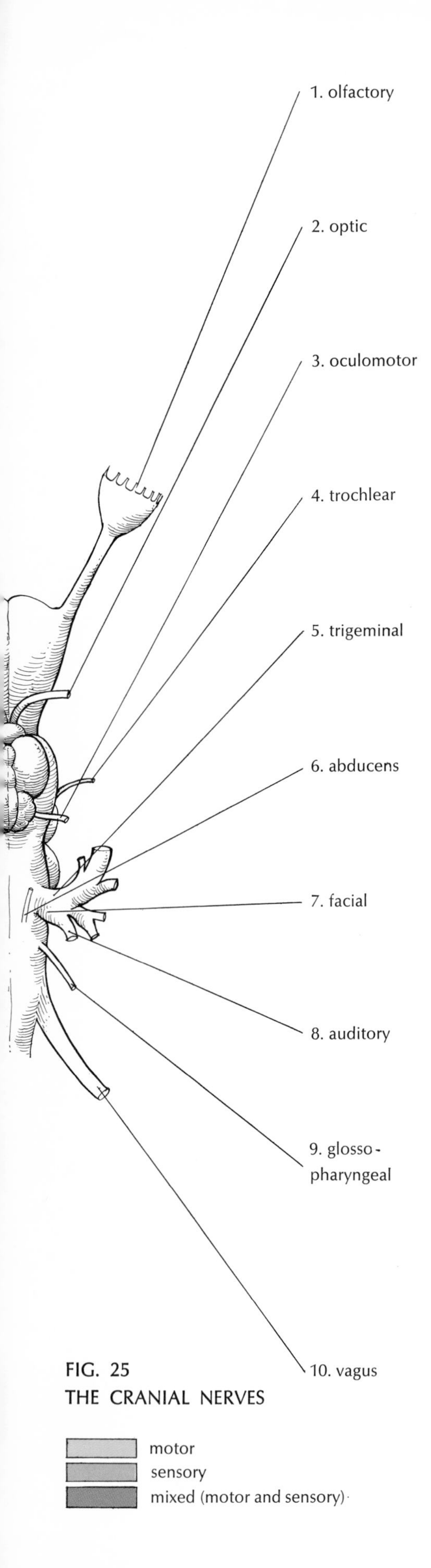


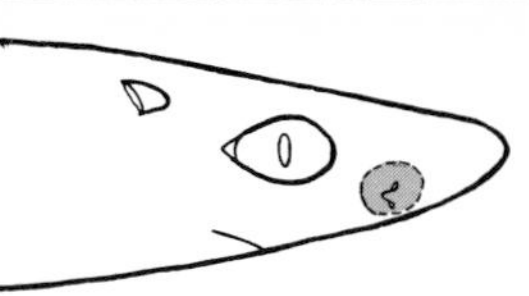

Sensory. From olfactory bulb to olfactory epithelium.

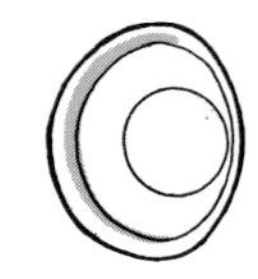

Sensory. From optic chiasm to retina.

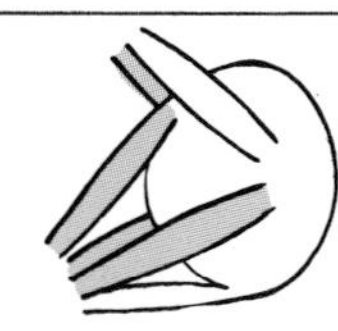

Motor. From floor of mesencephalon to superior rectus, medial rectus, inferior rectus, inferior oblique; & to smooth muscles of iris & ciliary body.

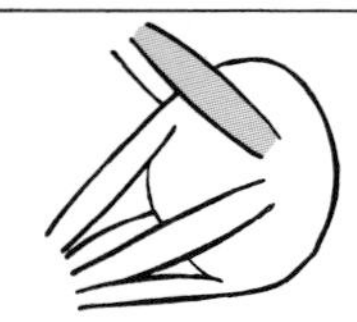

Motor. From roof of mesencephalon to superior oblique.

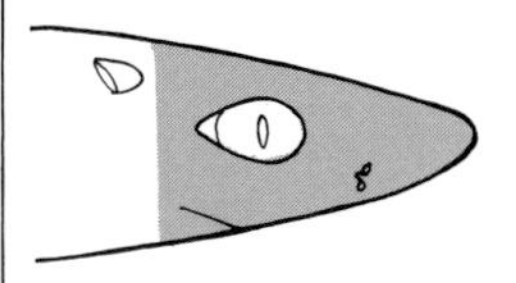

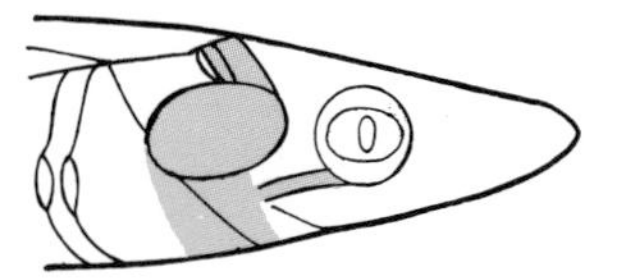

Mixed. From medulla, arising in common with 7 & 8.
Sensory to skin of head.
Motor to muscles of first gill arch.

Motor. From medulla to lateral rectus.

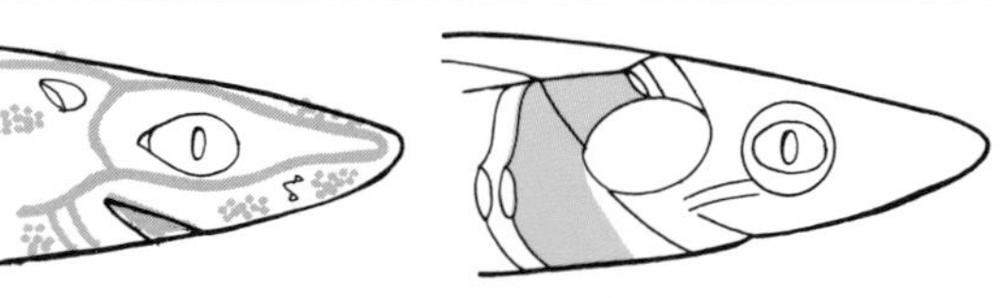

Mixed. From medulla, arising in common with 5 & 8.
Sensory to lateral line of head, ampullae of Lorenzini, and mouth.
Motor to muscles of second gill arch.

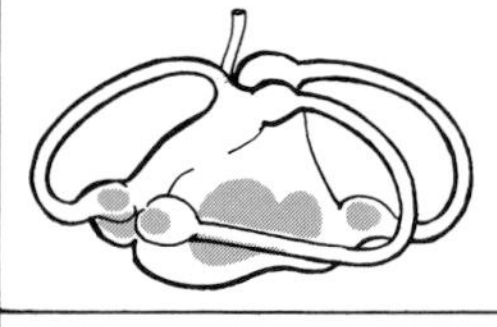

Sensory. From medulla, arising in common with 5 & 7, to sensory areas of membranous labyrinth.

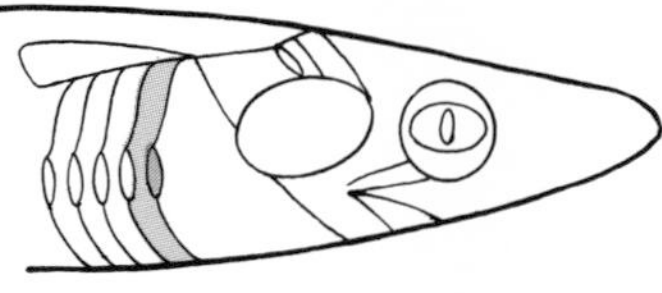

Mixed. From medulla.
Sensory to lateral line, first gill pouch, & pharynx.
Motor to muscles of third gill arch.

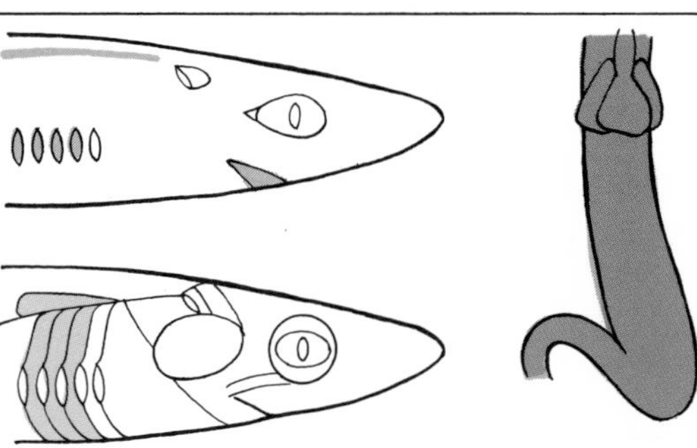

Mixed. From medulla.
Sensory to mouth, gill pouches 2-5, & lateral line.
Motor to muscles of gill arches 4-7 & cucullaris.
Mixed to heart & anterior part of alimentary canal.

FIG. 25
THE CRANIAL NERVES

- motor
- sensory
- mixed (motor and sensory)

0. The nervus terminalis is seen as a slender strand lying along the medial side of the olfactory tract, extending from the fissure between the two cerebral hemispheres to the olfactory bulb. Its function is not fully understood.

1. The olfactory nerve (sensory) consists of numerous neurons with cell bodies in the olfactory epithelium of the nasal sac. These neurons convey impulses from the olfactory epithelium to the olfactory bulb, from which they are relayed, via the olfactory tract, to the cerebral hemisphere.

2. The optic nerve (sensory) originates from the optic chiasm in the base of the diencephalon and passes laterally to the eye, where it pierces the sclera and the choroid to supply the retina.

3. The oculomotor nerve (motor) arises from the floor of the mesencephalon and emerges dorsal to the vascular sac. It enters the orbit near the origin of the superior rectus muscle, passing ventral to the superficial ophthalmic trunk, and divides into dorsal and ventral branches. The dorsal branch supplies the superior and medial rectus muscles. The ventral branch passes posteriorly around the inferior rectus muscle, which it supplies, and then extends anteriorly, crossing the ventral side of the inferior rectus, to supply the inferior oblique. The ventral branch also gives off a small ciliary nerve (difficult to identify), which conveys autonomic impulses to the smooth muscles of the iris and ciliary body, regulating accommodation and the size of the pupil.

4. The trochlear nerve (motor) originates from motor cells in the floor of the mesencephalon, but its fibers pass dorsally and it leaves the brain from the roof of the mesencephalon, emerging between the optic lobe and the cerebellum. It passes anteriorly, entering the orbit about half way between the origin of the superior rectus and the origin of the superior oblique, and crosses ventral to the superficial ophthalmic trunk to supply the superior oblique muscle.

Nerves 5, 7, and 8 originate together by a large common stem at the anterior end of the medulla, just ventral to the auricles of the cerebellum. The auditory nerve (8) branches from the posterior aspect of the common stem near its origin, and goes directly to the membranous labyrinth. After giving off the auditory nerve, the common stem, which now contains only fibers of the trigeminal (5) and facial (7) nerves, passes through a foramen in the posterior part of the medial wall of the orbit and divides into six branches. Two of these branches, the superficial ophthalmic trunk and the infraorbital trunk, contain components of both cranial nerves 5 and 7, but these components cannot be distinguished grossly. Two branches, the deep ophthalmic nerve and the mandibular nerve, are made up exclusively of fibers of the trigeminal nerve (5). The hyomandibular and palatine nerves consist exclusively of fibers of the facial nerve (7).

5. The trigeminal nerve (motor and sensory) has four branches: the superficial ophthalmic (part of the superficial ophthalmic trunk), the deep ophthalmic, the maxillary (part of the infraorbital trunk), and the mandibular.

The superficial ophthalmic branch of the trigeminal combines with a branch of the facial nerve to form the superficial ophthalmic trunk. The superficial ophthalmic trunk arises from the common stem just posterior to the origin of the rectus muscles and passes anteriorly along the medial wall of the orbit, lying dorsal to the superior rectus and superior oblique muscles. It gives off numerous branches, which pass through a series of foramina in the dorsal part of the orbit, after which it extends anteriorly into the rostrum, where it breaks up into many small branches. The trigeminal component of the superficial ophthalmic trunk conveys sensory impulses from the skin dorsal to the orbit.

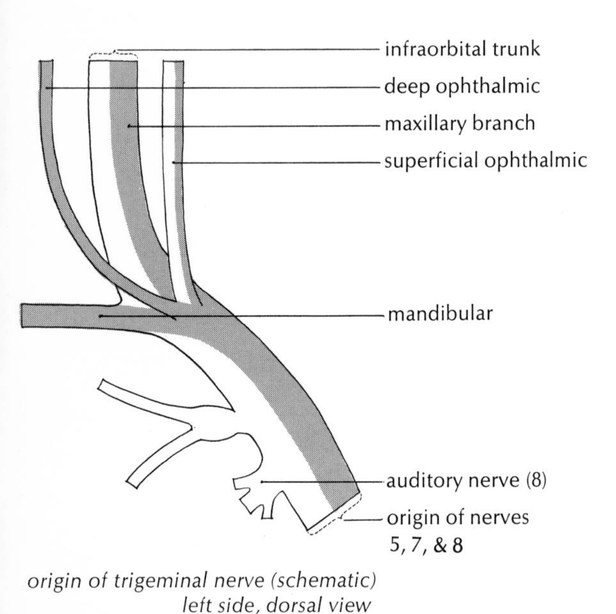

*origin of trigeminal nerve (schematic)*
*left side, dorsal view*

The deep ophthalmic nerve arises from the common stem near the origin of the superficial ophthalmic trunk and passes laterally, ventral to the superior rectus muscle, to the medial aspect of the eyeball, to which it gives a few minute branches. It then turns medially, passing ventral to the superior oblique muscle, makes a connection with the superficial ophthalmic trunk, and extends anteriorly, supplying sensory fibers to the rostrum.

The infraorbital trunk is a broad band composed of the maxillary branch of the trigeminal nerve and the buccal branch of the facial nerve. It passes anteriorly through the orbit, lying ventral to the inferior rectus and inferior oblique muscles, and divides into numerous branches on the ventral side of the rostrum. The trigeminal component of the infraorbital trunk consists of sensory fibers to the skin of the rostrum.

The mandibular nerve passes laterally between the eye and the otic capsule, and then turns ventrally to supply the muscles of the jaw. It is sensory to the skin of the lower jaw, and motor to the muscles of the first gill arch: the levator maxillae superioris, the craniomaxillaris, the preorbitalis, the quadratomandibularis, and the anterior part of the intermandibularis.

6. The abducens nerve (motor) originates from the ventral side of the medulla near the midline, just posterior to the common stem of nerves 5, 7, and 8. It passes anteriorly along the ventral side of the common stem to the lateral rectus muscle, which it innervates.

7. The facial nerve (motor and sensory) has four branches: the superficial ophthalmic (part of the superficial ophthalmic trunk), the buccal (part of the infraorbital trunk), the palatine, and the hyomandibular.

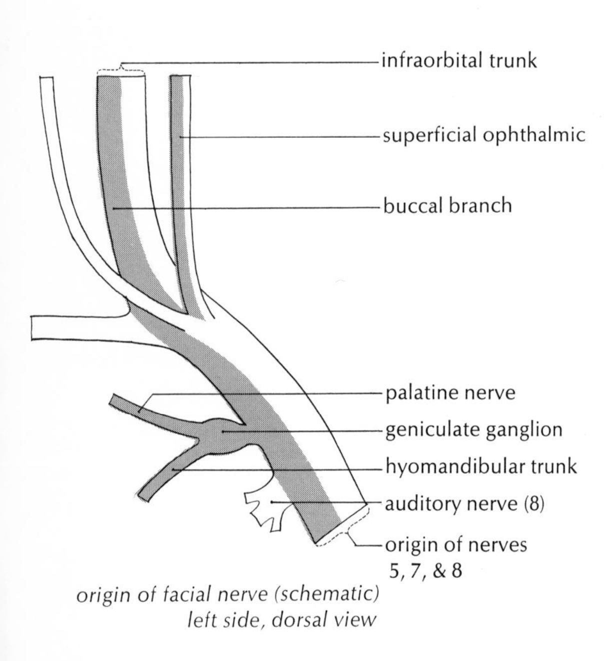

*origin of facial nerve (schematic)*
*left side, dorsal view*

The superficial ophthalmic branch of the facial nerve supplies sensory fibers to the lateral line canal and the ampullae of Lorenzini above the eye.

The buccal branch of the facial nerve and the maxillary branch of the trigeminal nerve together form the infraorbital trunk. The buccal branch of the facial nerve supplies sensory fibers to the lateral line canal and the ampullae of Lorenzini below the eye.

The hyomandibular nerve and the palatine nerve arise by a short common stem just anterior to the origin of the auditory nerve. This stem bears a ganglion, the geniculate ganglion, from which the palatine nerve and the hyomandibular trunk pass laterally. The palatine nerve conveys sensory impulses from the taste buds and the epithelium of the mouth. The hyomandibular trunk passes posteriorly around the otic capsule. It emerges dorsal to the spiracle and can be seen superficially in the lateral view of the head (Fig. 5, p. 10), lying between the quadratomandibularis and the epihyoideus muscles. The hyomandibular trunk conveys sensory impulses from the tongue and the floor of the mouth, and from the lateral line canals and the ampullae of Lorenzini in the region of the mouth. It conveys motor impulses to the muscles of the second (hyoid) gill arch: the epihyoideus, the hyoid levator, the second constrictor, the interhyoideus, and the posterior part of the intermandibularis.

8. The auditory nerve (sensory) arises from the posterior part of the common stem of nerves 5, 7, and 8 and enters the otic capsule, where it immediately divides into several branches to supply the sensory areas of the membranous labyrinth. Chip away the otic capsule on the right and trace its branches as illustrated in Figure 24. Sensory impulses from the membranous labyrinth are conveyed by the auditory nerve to the acousticolateral area of the medulla.

9. The glossopharyngeal nerve (motor and sensory) originates from the lateral aspect of the medulla posterior to the common stem of nerves 5, 7, and 8. It passes posteriorly, ventral to the otic capsule. Near the first gill pouch it bears a swelling termed the petrosal ganglion. Distal to the petrosal ganglion it divides into three main branches: pretrematic, post-trematic, and pharyngeal. The pretrematic branch passes ventrally anterior to the first gill pouch; the post-trematic branch passes ventrally posterior to the first gill pouch; and the pharyngeal branch goes to the pharynx. The pretrematic branch is sensory to the first gill pouch; the post-trematic branch is sensory to the first gill pouch and motor to the muscles of the third gill arch; the pharyngeal branch is sensory to the pharynx. The glossopharyngeal nerve also gives a small sensory branch (difficult to identify) to the lateral line.

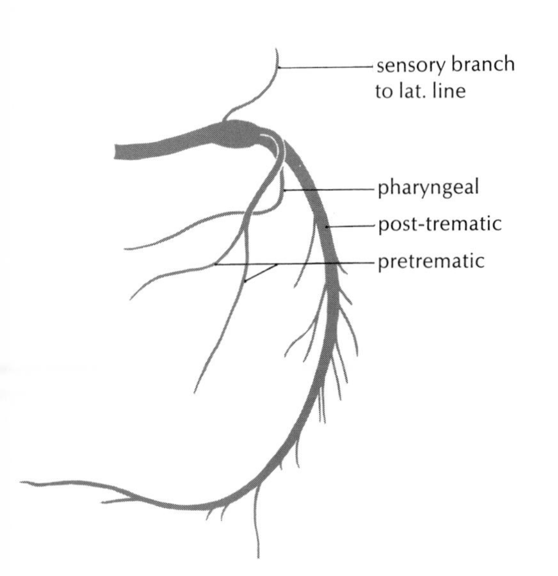


*glossopharyngeal nerve, left lateral view (after Norris & Hughes)*

See the vagus nerve, the hypobranchial nerve, and the first spinal nerves as illustrated in Figure 26. If a dissection of this area is attempted, begin by finding the branchial branches of the vagus nerve in the floor of the anterior cardinal sinus and trace them medially, picking away the epaxial muscles to expose the course of the vagus and the hypobranchial nerves. Shave away the dorsal part of the vertebral column, exposing the spinal cord, and trace the anterior spinal nerves as far as possible. The dorsal and ventral roots of the spinal nerves are delicate and difficult to identify.

10. The vagus nerve (motor and sensory) is a large nerve that originates by numerous rootlets from the lateral aspect of the medulla posterior to the origin of the glossopharyngeal nerve. It passes posteriorly medial to the gill pouches and enters the anterior cardinal sinus. Within the anterior cardinal sinus the vagus divides into a large visceral branch and a smaller lateral branch, which lies medial to the visceral branch. The lateral branch is sensory to the lateral line canal. The visceral branch gives off four branchial nerves to the remaining gill pouches. Each branchial nerve crosses the floor of the anterior cardinal sinus and divides into pretrematic, post-trematic, and pharyngeal branches, with function and distribution like the corresponding branches of the glossopharyngeal nerve (9). The visceral branch gives off a small motor branch to the cucullaris and extends posteriorly, carrying autonomic sensory and motor fibers to the heart and the anterior part of the alimentary canal.

The occipital nerves are two (sometimes three) small nerves that originate just posterior to the vagus. They are the ventral roots of spinal nerves, the dorsal roots of which have disappeared or been incorporated into the vagus. The occipital nerves and the first two spinal nerves form the hypobranchial nerve, which passes

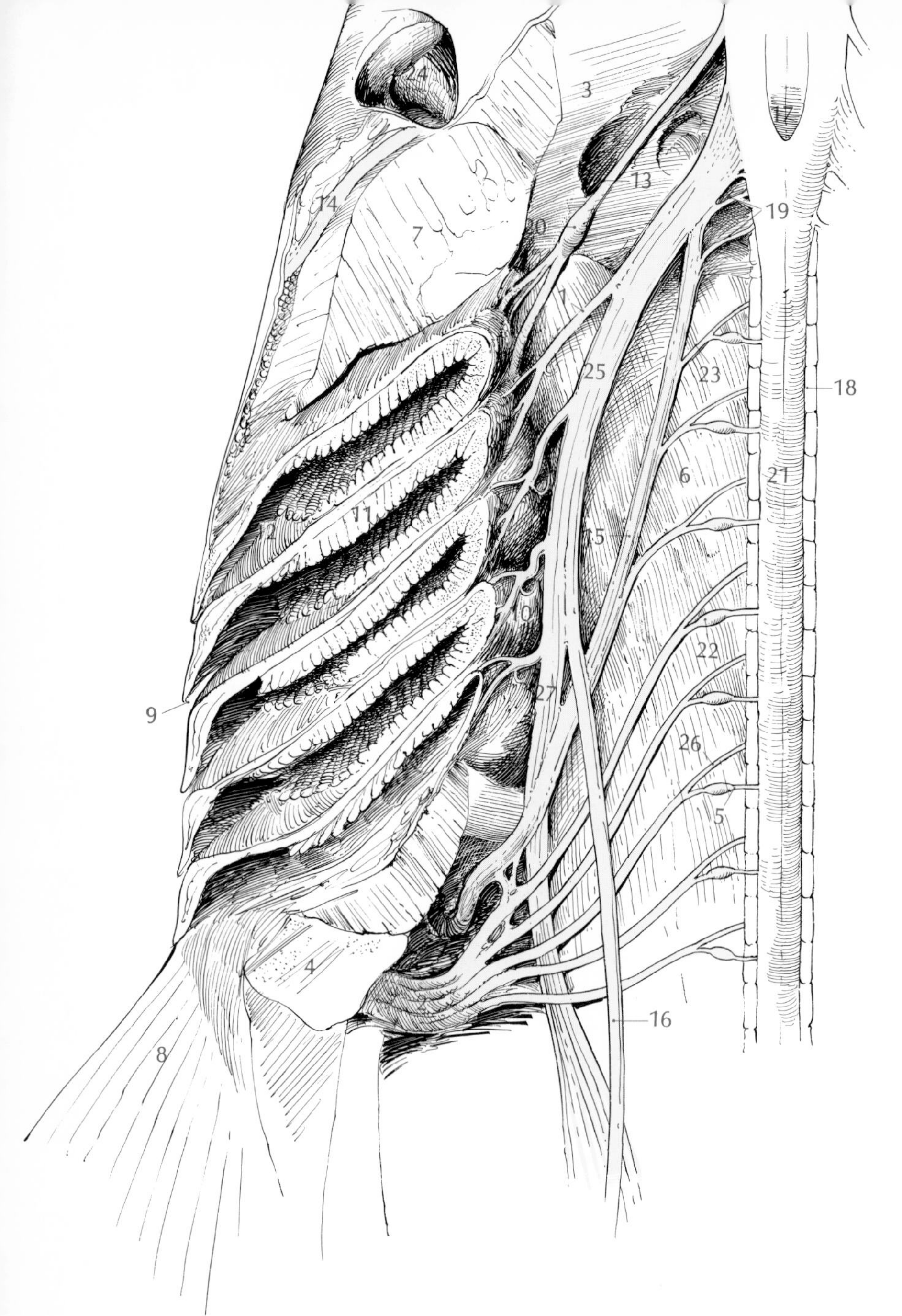


FIG. 26
THE VAGUS AND HYPOBRANCHIAL NERVES, DORSAL VIEW

1 branchial nerve
2 cervicobrachial plexus
3 chondrocranium
4 coracoid bar
5 dorsal root of spinal nerve
6 epaxial muscles
7 epihyoideus muscle
8 extensor muscle of pectoral fin
9 external gill slit
10 floor of anterior cardinal sinus
11 gill lamellae
12 gill pouch
13 glossopharyngeal nerve (9)
14 hyomandibular trunk (7)
15 hypobranchial nerve
16 lateral branch of vagus nerve
17 medulla
18 neural arch
19 occipital nerves
20 petrosal ganglion
21 spinal cord
22 spinal ganglion
23 spinal nerve
24 spiracle
25 vagus nerve (10)
26 ventral root of spinal nerve
27 visceral branch of vagus nerve

laterally, crossing the vagus about the level of the fifth gill pouch. After crossing the vagus the hypobranchial nerve turns ventrally and anteriorly. It is sensory to the skin ventral to the gill slits and motor to the hypobranchial muscles: the common coracoarcuals, the coracomandibular, the coracohyoid, and the coracobranchials.

The third spinal nerve appears to join the hypobranchial, but all its fibers separate from the hypobranchial near the shoulder girdle and join the cervicobrachial plexus.

Remove the brain and examine its lateral aspect, identifying the structures illustrated in Figure 27 (lateral view). Except for the optic tract and the habenula, all these structures have been seen in the dorsal and ventral views of the brain, with which the lateral view should be compared.

Now cut the brain in the sagittal plane and identify the structures illustrated in Figure 27 (sagittal section). Probe the foramen of Monro, observing that the choroid plexus extends into the lateral ventricle. Compare the ventricles as seen in the sagittal section with the view of the ventricles illustrated in the marginal diagram on page 38. Observe the boundaries of the third ventricle, and the extension of the third ventricle into the infundibulum.

The epithalamus, or roof of the third ventricle, includes several structures that are of significance in vertebrate evolution, but they are difficult to identify in the adult. These structures can best be seen in a sagittal section of the embryonic brain, as illustrated in the marginal diagram. At this stage of development it can be seen that the structures of the epithalamus are derived from a membrane that originally formed a single continuous sheet over the third ventricle. Anteriorly the roof of the third ventricle forms a thin-walled pouch, the paraphysis, which extends forward in the midline over the cerebral hemispheres. It is part of the telencephalon. Posterior to the paraphysis the roof of the third ventricle invaginates to form the velum transversum, a membranous partition that forms the boundary between the diencephalon and the telencephalon. Posterior to this is a small mass of nervous tissue termed the habenula, from which a long, slender process, the pineal body, extends anteriorly. Behind the pineal body is the posterior commissure, a fiber tract connecting opposite sides of the brain.

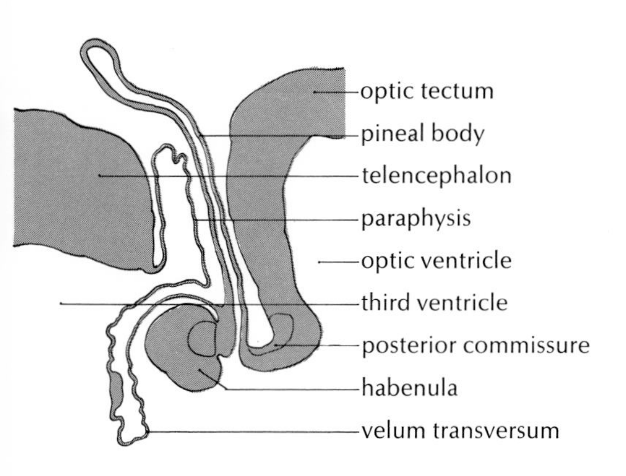


*epithalamus of 86 mm. embryo (after Minot)*

The thalamus is an area of gray matter in the lateral wall of the third ventricle.

Compare the components of the hypothalamus in the ventral view and in the sagittal section. The mammillary body, which was not visible in the ventral view, can be seen in the sagittal section. Also observe the tegmentum, or floor of the mesencephalon. Pass a probe from the aqueduct into the optic ventricle. In the lateral wall of the fourth ventricle observe the acousticolateral area, the somatic sensory column, and the visceral sensory column, described on page 42.

The spinal cord is a slightly flattened tube of fairly uniform diameter, lying within the neural canal. It is continuous anteriorly with the medulla, and posteriorly it extends almost to the tip of the tail. In cross section it can be seen that its shape is oval, and that it contains a small central canal, which is continuous anteriorly with the fourth ventricle of the medulla, and extends throughout the length of the spinal cord. The spinal cord is surrounded by a membrane termed the primitive meninx, which is continuous with the primitive meninx of the brain.

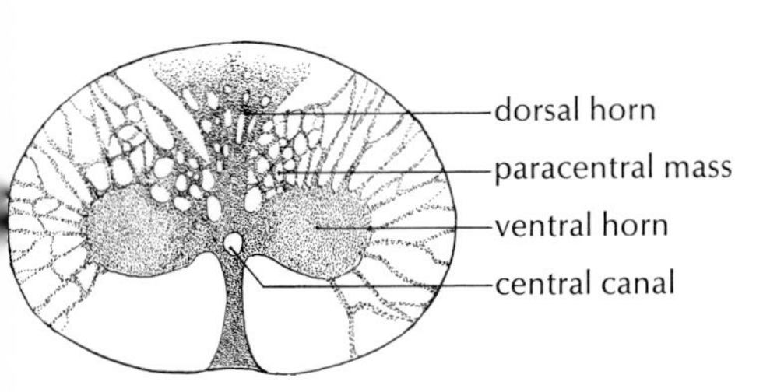


*cross section of spinal cord (after Daniel, from Sterzi)*

In a stained section it can be seen that the spinal cord consists of a central area of gray matter, consisting chiefly of cell bodies, and a peripheral area of white matter, consisting chiefly of fiber tracts.

A series of spinal nerves arise from the spinal cord at segmental intervals, so that each nerve corresponds to a vertebral body and a myotome. Each nerve is attached to the spinal cord by a dorsal (sensory) and a ventral (motor) root, which exit from the neural canal through foramina in the neural arch. The origins of the dorsal and ventral roots are staggered, so that the ventral root of each nerve arises somewhat anterior to the origin of the corresponding dorsal root. Outside the neural canal the dorsal root bears a spinal ganglion, containing the cell bodies of sensory fibers. Near the union of the dorsal and ventral roots several branches, too small to see in gross dissection, are given off to the nearby muscles.

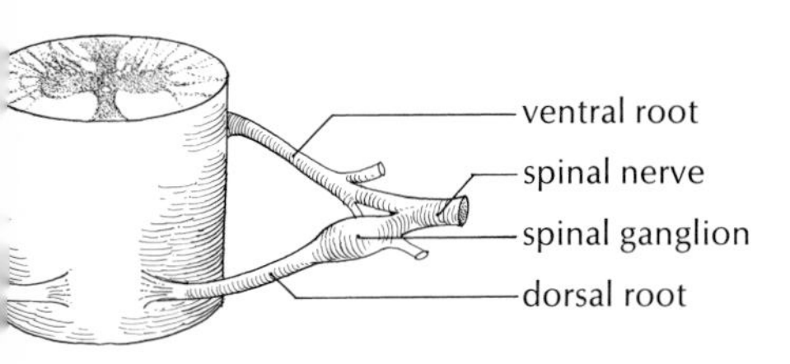


*spinal cord, dorsal view*

The first two spinal nerves contribute to the hypobranchial nerve. Spinal nerves 3 through 14 supply the pectoral fin, and the first four or five of these are connected to form the cervicobrachial plexus. The pelvic fin is supplied by ten nerves, of which the last four or five are connected to form the pelvic plexus.

In the trunk region the epaxial part of each myotome is supplied by the corresponding spinal nerve, but as they pass ventrally the spinal nerves fuse to form a plexus, so that the region innervated by a single spinal nerve extends beyond the boundaries of the corresponding myotome in the hypaxial region. In the tail there is a more primitive arrangement: the dorsal and ventral roots of each spinal nerve do not fuse, but remain separate.

Expose the dorsal wall of the posterior cardinal sinus. The nerves of the pectoral fin will be seen as white bands extending laterally toward the fin. The spinal nerves of the trunk can be seen on the inner body wall; they are delicate strands that lie along the myosepta between the myotomes. By blunt dissection carefully separate the extensor of the pelvic fin from the hypaxial muscles to which it is attached in order to see the nerves of the pelvic fin. The nerves run from the body wall to the fin, embedded in connective tissue.

It is not practical to demonstrate the autonomic system in the dogfish by gross dissection. In the head the autonomic components are the ciliary ganglia and plexus, connected with the oculomotor nerve, and ganglia associated with the post-trematic branches of the branchial nerves. In the trunk there are ganglia in the dorsal wall of the posterior cardinal sinus and along the medial borders of the kidneys, in many cases closely associated with the scattered tissue that corresponds to the adrenal medulla of mammals. In some cases two or more of these ganglia are connected, but they do not form a complete chain like the sympathetic chain of higher vertebrates. They receive preganglionic fibers from the spinal cord and send postganglionic fibers to the kidneys and the posterior part of the alimentary canal. The double innervation of organs by antagonistic sympathetic and parasympathetic fibers is not found in the dogfish.

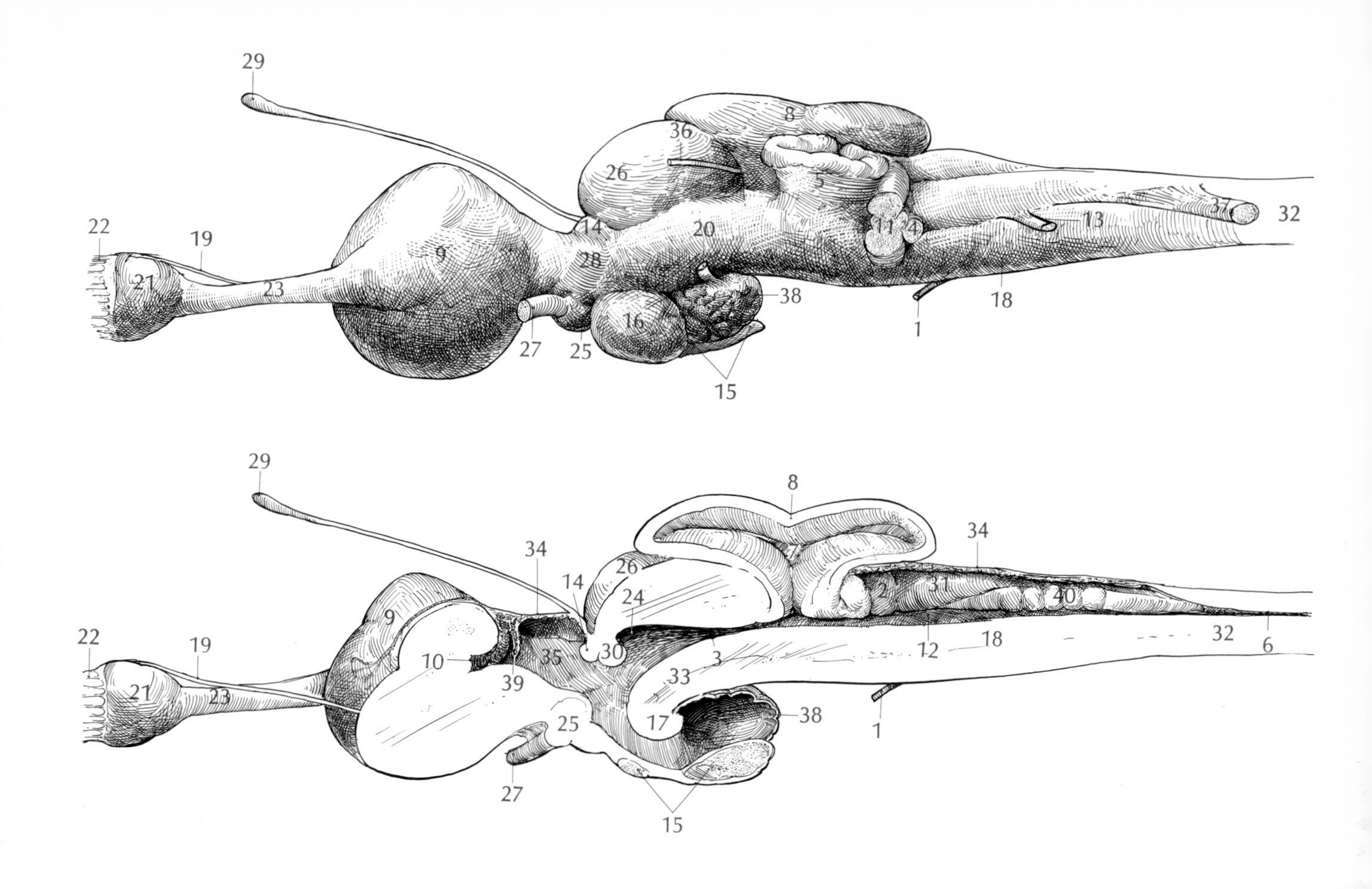


FIG. 27
LATERAL VIEW AND SAGITTAL SECTION OF THE BRAIN

1 abducens nerve (6)
2 acousticolateral area
3 aqueduct
4 auditory nerve (8)
5 auricle of cerebellum
6 central canal
7 cerebellar ventricle
8 cerebellum
9 cerebral hemisphere
10 choroid plexus in foramen of Monro
11 common origin of trigeminal (5) and facial (7) nerves
12 fourth ventricle
13 glossopharyngeal nerve (9)
14 habenula
15 hypophysis
16 inferior lobe of infundibulum
17 mammillary body
18 medulla
19 nervus terminalis (0)
20 oculomotor nerve (3)
21 olfactory bulb
22 olfactory nerve (1)
23 olfactory tract
24 opening into optic ventricle
25 optic chiasm
26 optic lobe
27 optic nerve (2)
28 optic tract
29 pineal body
30 posterior commissure
31 somatic sensory column
32 spinal cord
33 tegmentum
34 tela choroidea
35 third ventricle and thalamus
36 trochlear nerve (4)
37 vagus nerve (10)
38 vascular sac
39 velum transversum
40 visceral sensory column

# THE SENSE ORGANS

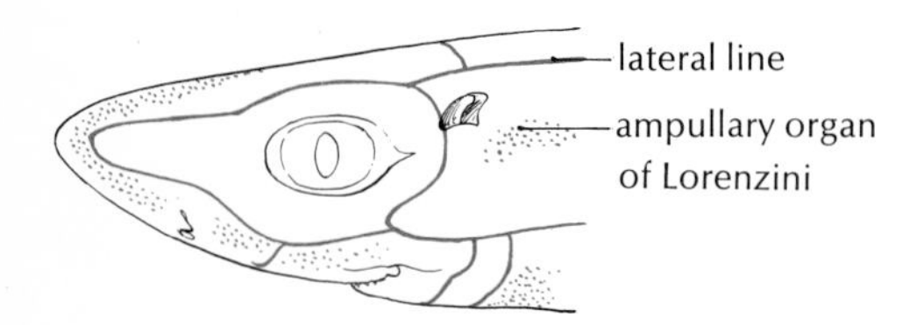


The lateral line system, found only in fishes and larval amphibians, is a system of canals and sensory cells capable of responding to disturbances in the water. This system enables the fish to detect the presence of nearby objects that give off waves or reflect waves created by the fish itself. The lateral line canals are a series of interconnected canals on the head and a single canal that runs the length of the body on either side. These canals lie just under the skin and are connected with the surface at intervals by branch canals which admit water. Ciliated sensory cells termed neuromasts, which register water vibrations, lie within the canal. In addition to the neuromasts in the lateral line canals, there are neuromasts in separate pits. These structures are termed pit organs, and are located near the base of the pectoral fin and the gill slits. The lateral line system is innervated by branches of cranial nerves 7, 9, and 10.

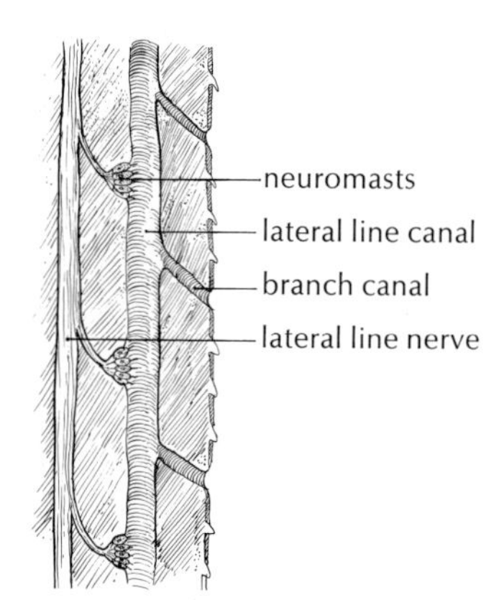


*longitudinal section of lateral line canal (after Goodrich)*

Press the skin on the top of the head between the eyes and observe that it is perforated by numerous pores which exude a gelatinous secretion. Similar pores can be seen posterior to the eyes and on the ventral surface of the snout. Remove an area of skin from the top of the head and observe that beneath the pores are numerous small tubes. These are the canals of Lorenzini. At the end opposite the pore, each canal bears a small sensory vesicle, the ampulla of Lorenzini, which is divided into numerous secondary vesicles. These vesicles are innervated by branches of the facial nerve. The ampullae of Lorenzini respond to changes in temperature, water pressure, weak electric fields, and salinity, but their specific function in the life of the dogfish is not fully understood.

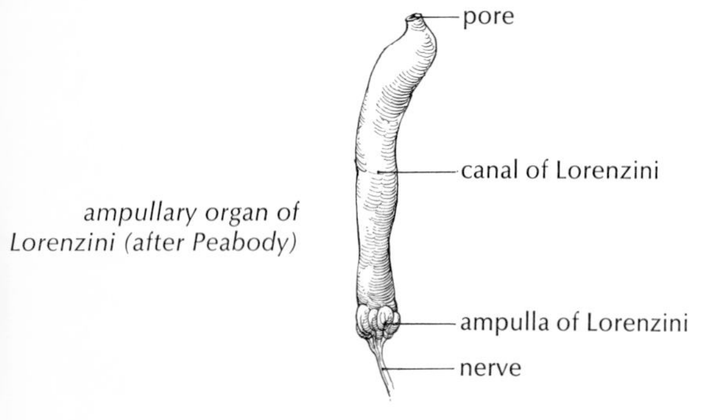


*ampullary organ of Lorenzini (after Peabody)*

The olfactory sacs are roughly spherical structures containing a series of radial folds termed olfactory lamellae. Each sac communicates with the water via an opening, termed the external naris, on the ventral surface of the head. The cavity of the nasal sac has no communication with the oral cavity. The external naris is protected by a flap of skin that allows water to circulate through the olfactory sac. The olfactory lamellae are covered by olfactory epithelium, within which are neurosensory cells. These cells send numerous short fibers to the olfactory bulb, where they synapse with other fibers that pass through the olfactory tract to the cerebral hemisphere. Observe

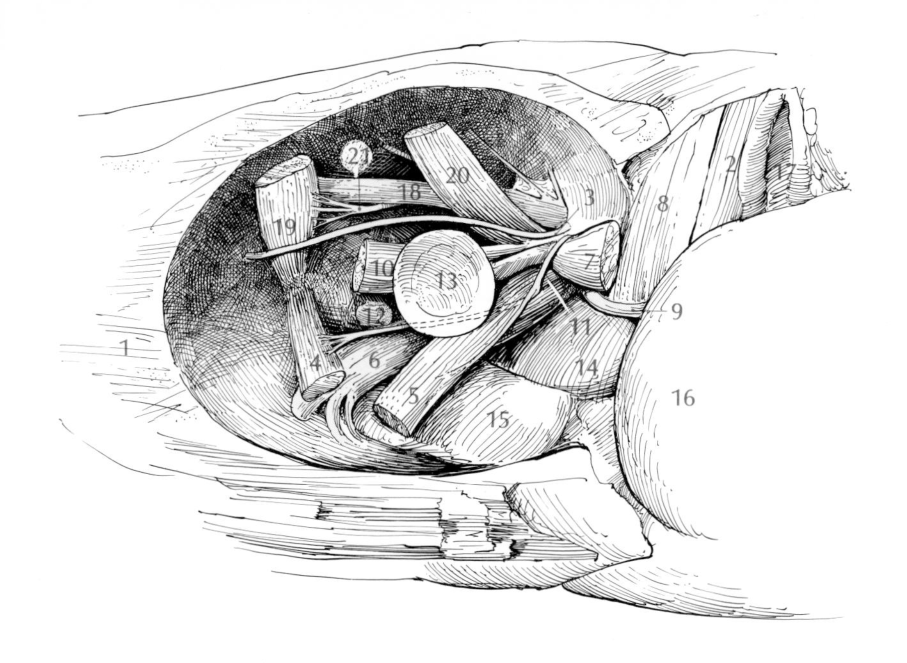


FIG. 28
THE CONTENTS OF THE LEFT ORBIT AFTER REMOVAL OF THE EYE

1 chondrocranium
2 craniomaxillaris
3 deep ophthalmic nerve (5)
4 inferior oblique
5 inferior rectus
6 infraorbital trunk (5, 7)
7 lateral rectus
8 levator maxillae superioris
9 mandibular nerve (5)
10 medial rectus
11 oculomotor nerve (3)
12 optic nerve (2)
13 optic pedicle
14 palatoquadrate cartilage
15 preorbitalis
16 quadratomandibularis
17 spiracle
18 superficial ophthalmic trunk (5, 7)
19 superior oblique
20 superior rectus
21 trochlear nerve (4)

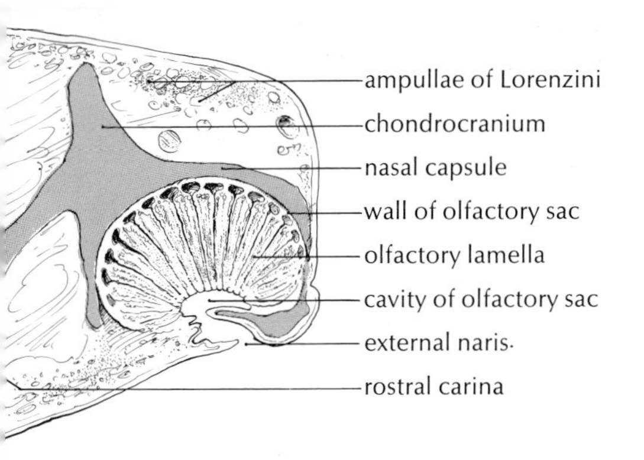


*cross section of olfactory sac and nasal capsule*

the olfactory sacs in the dorsal and ventral views of the head illustrated in Figures 23 and 24, and make a cross section of one of the sacs as illustrated in the marginal diagram.

Observe the muscles of the eye as illustrated in Figures 23, 24, 28, and 29. The eye is moved by six muscles: four rectus muscles and two oblique muscles. The four rectus muscles originate near the posterior end of the orbit, near the origin of the optic pedicle, and pass forward to insert on the eyeball. They are named according to their position: the medial rectus lies between the medial wall of the orbit and the eyeball; the lateral rectus lies on the posterior side of the eyeball; the superior rectus lies above the eye; and the inferior rectus lies below the eye. The superior and inferior oblique muscles originate from the anterior end of the orbit and pass posteriorly to insert on the eyeball. The superior oblique is innervated by the trochlear nerve (4). The lateral rectus is innervated by the abducens nerve (6). The other muscles of the eye are innervated by the oculomotor nerve (3).

The optic pedicle is a cartilaginous structure consisting of a flattened stalk bearing a disk on its lateral end. It originates between the origins of the rectus muscles and serves as a ball-and-socket support for the eyeball, which is also supported and cushioned by gelatinous material in the orbit.

If you made your dissection of the head from the dorsal aspect, remove the superior oblique and the superior rectus muscles. If you made your dissection of the head from the inferior aspect, remove the inferior oblique and inferior rectus muscles. Cut the eyeball as illustrated in Figure 29.

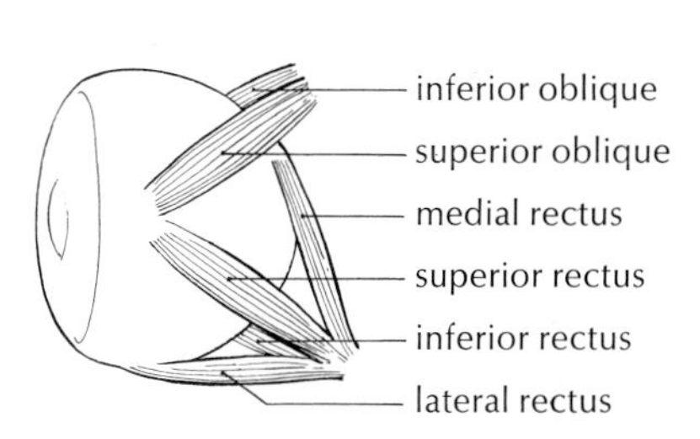


*left eye, dorsal view*

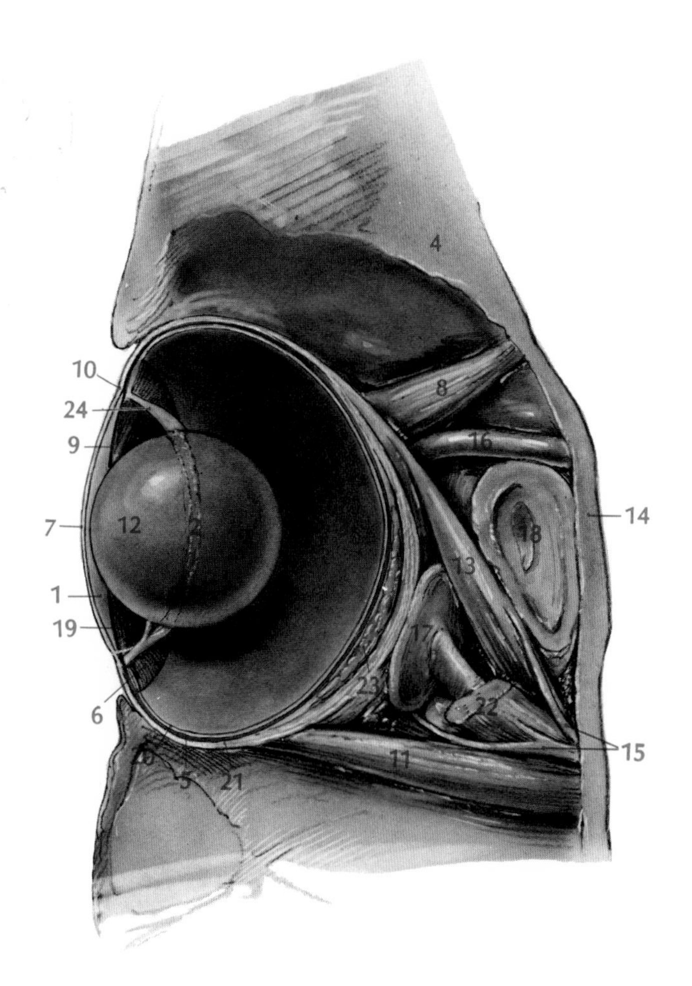


FIG. 29
DORSAL VIEW OF THE LEFT EYE

The dorsal part of the eyeball, and the superior rectus and superior oblique muscles are removed.

1 anterior chamber
2 attachment of suspensory membrane of the lens (cut)
3 cavity of vitreous humor
4 chondrocranium
5 choroid
6 ciliary body
7 cornea
8 inferior oblique
9 iris
10 junction of cornea and sclera
11 lateral rectus
12 lens
13 medial rectus
14 medial wall of orbit
15 oculomotor nerve (3)
16 optic nerve (2)
17 optic pedicle
18 orbital process of palatoquadrate cartilage
19 posterior chamber
20 retina
21 sclera
22 superior rectus
23 suprachoroidea
24 suspensory membrane of lens

The upper and lower eyelids are immovable folds of skin. A thin membrane, the conjunctiva, extends from the lid onto the eye, covering its exposed surface. Find the conjunctiva by probing at the point where the lid meets the eyeball.

The eye consists of three layers or tunics: the external tunic, the middle tunic, and the inner tunic.

The external tunic includes the sclera and the cornea. The sclera is a tough protective covering consisting of white fibrous connective tissue; it makes up the medial part of the external tunic. Laterally it is continuous with the cornea, which is transparent and admits light to the eye. Observe that the medial part of the sclera, near the point where it is supported by the optic pedicle, is the thickest.

The middle tunic is heavily pigmented and appears black in dissection. It is the only one of the three layers of the eye that contains blood vessels in the adult. It consists of the choroid, the ciliary body, and the iris. The choroid is a vascular layer that lies between the retina and the sclera. The ciliary body is a thin, amuscular band, marked by faint radial lines, which forms the part of the middle tunic between the iris and the choroid. The iris is the anterior extension of the middle tunic, lying between the cornea and the lens. It is perforated by a central opening, the pupil, which controls the amount of light admitted to the eye. Medially the choroid is thickened by an area of connective tissue, veins, and lymph spaces which form an unpigmented layer, termed the suprachoroidea, at the point where the optic pedicle supports the eye.

The inner tunic of the eye is the retina. In dissection it appears as a delicate white membrane that can easily be detached from the choroid. Posteriorly it is continuous with the optic nerve; anteriorly it extends as a thin layer onto the medial surface of the ciliary body and iris.

The lens, which is almost spherical, is transparent in life. It is suspended in the cavity of the eyeball by a gelatinous suspensory membrane which originates from the ciliary body and attaches to the equator of the lens. Walls describes a dorsal thickening of this membrane, the dorsal suspensory ligament, and a ventral papilla, containing smooth muscle fibers, that extends from the ciliary body to the lens. This muscle, termed the protractor lentis, effects accommodation by pulling the lens toward the cornea. At rest the eye of the dogfish is focused on distant objects; when the lens is pulled toward the cornea, images of near objects come into focus.

The cavity of the eye posterior to the lens is termed the cavity of the vitreous humor. It is filled with a gelatinous substance termed the vitreous humor. The space anterior to the iris is termed the anterior chamber; the space between the iris and the suspensory membrane of the lens is termed the posterior chamber. The anterior and posterior chambers are filled with a clear liquid termed the aqueous humor.

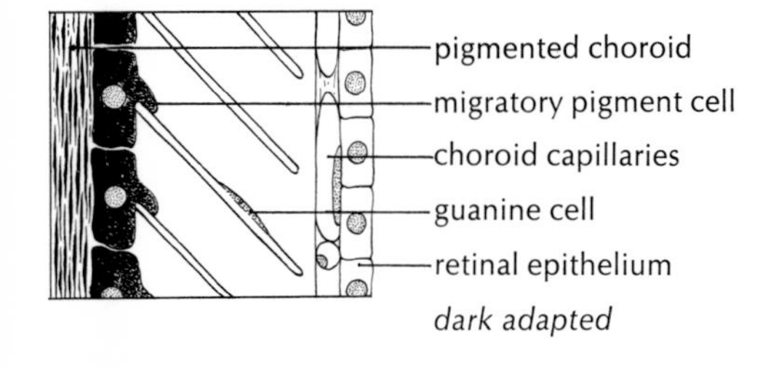


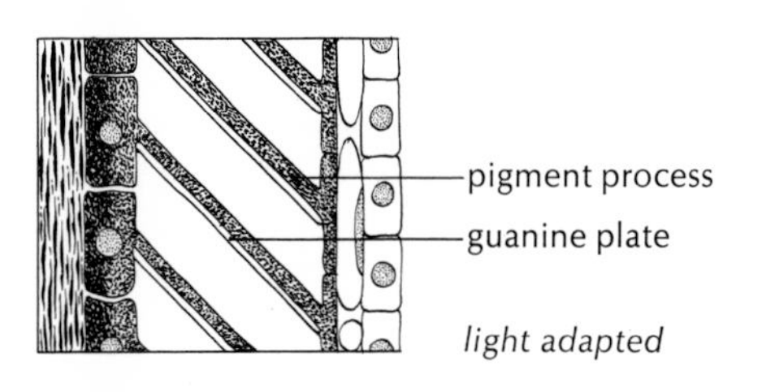


*section through choroid of* Mustelus mustelus *(after Walls, from Franz)*

The tapetum lucidum is an adaptation that enhances visual acuity in dim light. Light entering the eye passes through the retina; it is then reflected back through the retina a second time by the tapetum lucidum, thus doubling the apparent brightness of the light.

The elasmobranch tapetum is unique in that it can be covered by migratory processes from pigment cells in the choroid. When the eye is adapted to vision in the dark, light entering the eye passes through the retina and is reflected back by a series of flat overlapping guanin-filled cells that lie between the pigmented choroid and the retina. When the eye is adapted to light, migratory pigment processes from the cells of the choroid cover the guanin-filled cells, and the retina absorbs the light passing through instead of reflecting it.

The membranous labyrinth is a system of chambers and ducts within a cavity in the otic capsule. It is innervated by the auditory nerve, and serves to register sound and body position. The position of the membranous labyrinth can best be seen by examining a chondrocranium in which the cavity containing the membranous labyrinth has been filled with a colored mass.

Examine the intact head. In the midline between the spiracles there are two small openings in the skin. If you have difficulty finding these openings, scrape the skin with a knife. These are the external openings of the endolymphatic ducts, which form a communication between the membranous labyrinth and the water. Remove the skin and observe the endolymphatic duct leading toward the inner ear. See the dorsal view of the skull (Fig. 1, p. 2). The endolymphatic duct communicates with the membranous labyrinth via the endolymphatic foramen in the endolymphatic fossa. Embryologically the membranous labyrinth originates by invagination from the same placode that gives origin to the lateral line canals. The endolymphatic duct is a vestige of the original connection of the membranous labyrinth with the surface of the body.

In Figures 23 and 24 the membranous labyrinths are illustrated schematically. Because of their delicate structure, it is difficult to dissect them out of the surrounding cartilage. The position of the sacs and canals composing the inner ear can best be determined by shaving away the otic capsule with a razor blade.

The large central chamber to which the endolymphatic duct leads is the sacculus. A smaller chamber, the utriculus, is fused with the anterior side of the sacculus. Connected to the utriculus are two thin curved tubes: the anterior and horizontal semicircular ducts. Another curved tube, the posterior semicircular duct, is attached to the posterior side of the sacculus. The small protrusion at the posterior end of the sacculus is the lagena. At the ventral end of each semicircular duct is an ampulla, within which is a small sensory area termed a crista. The sacculus and the utriculus contain larger sensory areas termed maculae. Branches of the auditory nerve go to the cristae and the maculae. The sacculus contains a large mass of calcareous grains termed an otolith, and the utriculus contains a smaller otolith.

The cavity of the otic capsule within which the membranous labyrinth lies is termed the cartilaginous labyrinth. Strands of connective tissue extend between the membranous labyrinth and the walls of the cartilaginous labyrinth. The space between the membranous labyrinth and the cartilaginous labyrinth is filled with a clear fluid, the perilymph. The membranous labyrinth is filled with a similar fluid termed the endolymph.

The membranous labyrinth functions in the maintenance of muscle tone, in the perception of angular acceleration and static position, and in the perception of sound. Sound perception has been demonstrated in many bony fishes, but it is difficult to establish conditioned responses in elasmobranchs, and little is known about their hearing.

FIG. 30
THE LEFT MEMBRANOUS LABYRINTH
(after Marinelli and Strenger)

1 ampulla
2 anterior semicircular duct
3 auditory nerve (8)
4 crista
5 endolymphatic duct
6 horizontal semicircular duct
7 lagena
8 otolith
9 posterior semicircular duct
10 sacculus
11 utriculus

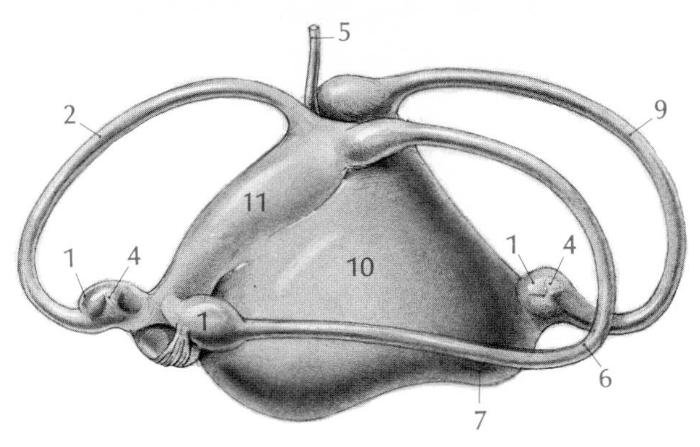

lateral view

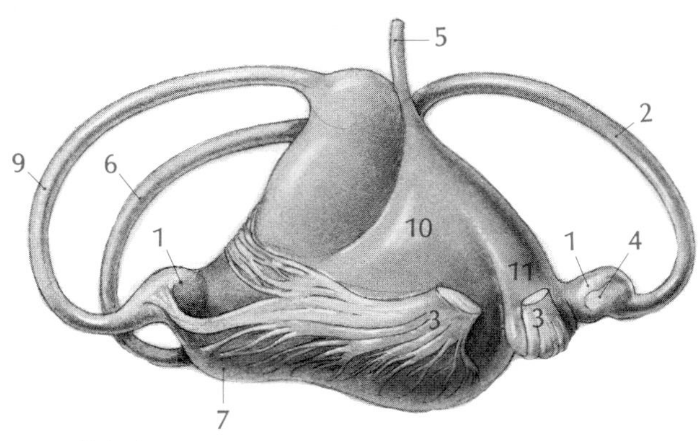

medial view

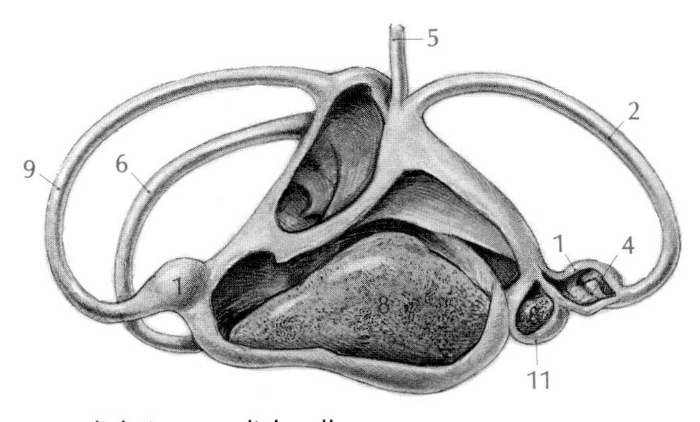

medial view, medial wall partially cut away

# BIBLIOGRAPHY

Brown, Margaret E. *The Physiology of Fishes.* New York: Academic Press, 1957.

Daniel, J. F. *The Elasmobranch Fishes.* 3d ed. Berkeley: University of California Press, 1934.

Goodrich, Stephen Edwin. *Studies on the Structure and Development of Vertebrates.* New York: Dover, 1958.

Hisaw, Frederick L., and A. Albert. "Observations on the Reproduction of the Spiny Dogfish, *Squalus Acanthias,*" *Biological Bulletin* 92 (1947):187-99.

Howell, A. Brazier. "The Architecture of the Pectoral Appendage of the Dogfish," *Journal of Morphology* 54, no. 2 (1933):399-413.

Hyman, Libbie H. *Comparative Vertebrate Anatomy.* 2d ed. Chicago: University of Chicago Press, 1942.

Marinelli, W., and A. Strenger. *Vergleichende Anatomie und Morphologie der Wirbeltiere.* III Lieferung. Wien: Franz Deuticke, 1959. This is the most complete and best illustrated account of the anatomy of the dogfish.

Marion, Guy Ellwood. "Mandibular and Pharyngeal Muscles of *Acanthias* and *Raia,*" *American Naturalist* 39 (1905):891-924.

Minot, Charles Sedgwick. "On the Morphology of the Pineal Region, Based upon Its Development in *Acanthias,*" *American Journal of Anatomy* 1 (1901-2):81-98.

Norris, H. W., and Sally P. Hughes. "The Cranial, Occipital, and Anterior Spinal Nerves of the Dogfish, *Squalus Acanthias,*" *Journal of Comparative Neurology* 31 (1919-20):293-402.

O'Donoghue, Charles H., and Eileen Bulman Abbott. "The Blood Vascular System of the Spiny Dogfish, *Squalus Acanthias,*" *Transactions of the Royal Society of Edinburgh* 55 (1927-28):823-90.

Walls, Gordon Lynn. *The Vertebrate Eye and Its Adaptive Radiation.* New York: Hafner, 1963.